Characteriz[illegible]g the U.S. Industrial Base for Coal-Powered Electricity

Constantine Samaras, Jeffrey A. Drezner, Henry H. Willis, Evan Bloom

Sponsored by the National Energy Technology Laboratory

Environment, Energy, and Economic Development

A RAND INFRASTRUCTURE, SAFETY, AND ENVIRONMENT PROGRAM

This research was sponsored by the National Energy Technology Laboratory and was conducted in the Environment, Energy, and Economic Development Program within RAND Infrastructure, Safety, and Environment.

Library of Congress Control Number: 2011940667

ISBN: 978-0-8330-5918-5

Cover image courtesy of Basin Electric Power Cooperative. Used with permission.

Published 2011 by the RAND Corporation
1776 Main Street, P.O. Box 2138, Santa Monica, CA 90407-2138
1200 South Hayes Street, Arlington, VA 22202-5050
4570 Fifth Avenue, Suite 600, Pittsburgh, PA 15213-2665
RAND URL: http://www.rand.org/
To order RAND documents or to obtain additional information, contact
Distribution Services: Telephone: (310) 451-7002;
Fax: (310) 451-6915; Email: order@rand.org

Preface

Coal-fired power plants provide nearly half of the electricity generated in the United States, yet most of the existing coal-fired power plant fleet is 25–45 years old. Deploying advanced coal-fired power plants and maintaining the existing fleet requires the capacity to provide equipment, skilled labor, and project management expertise. However, in the near term, the domestic industrial base to build future coal-based power plants faces several challenges, including low demand for new conventional and advanced coal-fired units over the next few years, regulatory uncertainty regarding reducing conventional pollutants and greenhouse gases, competition from other fuels, and an aging workforce.

To better understand these challenges, The National Energy Technology Laboratory (NETL) asked the RAND Corporation to describe the current state of the domestic industrial base for coal-based electricity generation. RAND was asked to focus on coal-fired electricity-generation unit design and construction, technological development, and equipment manufacturing. This monograph describes the findings from this effort. It should be of interest to decisionmakers and analysts addressing issues associated with U.S. capabilities to design, manufacture, construct, and maintain coal-fired electricity-generation units. The analysis should also be of interest to plant owners, equipment manufacturers, labor organizations, and construction firms interested in the current state of the domestic industrial base to generate electricity from coal. The research and analysis reported here builds on prior work RAND has performed in both energy and industrial-base analyses, particularly the following:

- David S. Ortiz, Aimee E. Curtright, Constantine Samaras, Aviva Litovitz, and Nicholas Burger, *Near-Term Opportunities for Integrating Biomass into the U.S. Electricity Supply: Technical Considerations*, Santa Monica, Calif.: RAND Corporation, TR-984-NETL, 2011
- Somi Seong, Obaid Younossi, Benjamin W. Goldsmith, Thomas Lang, and Michael J. Neumann, *Titanium: Industrial Base, Price Trends, and Technology Initiatives*, Santa Monica, Calif.: RAND Corporation, MG-789-AF, 2009
- John F. Schank, Mark V. Arena, Paul DeLuca, Jessie Riposo, Kimberly Curry Hall, Todd Weeks, and James Chiesa, *Sustaining U.S. Nuclear Submarine Design Capabilities*, Santa Monica, Calif.: RAND Corporation, MG-608-NAVY, 2007.

The RAND Environment, Energy, and Economic Development Program

This research was conducted in the Environment, Energy, and Economic Development Program (EEED) within RAND Infrastructure, Safety, and Environment (ISE). The mission of ISE is to improve the development, operation, use, and protection of society's essential physical assets and natural resources and to enhance the related social assets of safety and security of individuals in transit and in their workplaces and communities. The EEED research portfolio addresses environmental quality and regulation, energy resources and systems, water resources and systems, climate, natural hazards and disasters, and economic development—both domestically and internationally. EEED research is conducted for government, foundations, and the private sector.

Questions or comments about this monograph should be sent to the project leaders, Jeffrey Drezner (Jeffrey_Drezner@rand.org) and Henry Willis (Henry_Willis@rand.org). Information about EEED is available online (http://www.rand.org/ise/environ). Inquiries about EEED should be sent to the following address:

Keith Crane, Director
Environment, Energy, and Economic Development Program, ISE
RAND Corporation
1200 South Hayes Street
Arlington, VA 22202-5050
703-413-1100, x5520
Keith_Crane@rand.org

Contents

APPENDIXES

Figures

Tables

Summary

Coal-fired generating units provide approximately 46 percent of the electricity generated in the United States, yet most of the existing coal-fired electricity-generating fleet is 25–45 years old (Energy Information Administration [EIA], 2011c; Ventyx, 2011). Deploying new coal-fired electricity-generating units (EGUs) and maintenance of the existing fleet requires an industrial capacity to provide equipment, skilled labor, and project management expertise. However, in the near term, the domestic industrial base to provide the capacity for future coal-based EGUs faces several challenges. These include low demand for new conventional and advanced coal units in the next several years, regulatory uncertainty regarding emission standards for conventional pollutants and greenhouse gases, competition from other electricity-generation fuels, and an aging workforce.

To better understand these challenges, NETL asked RAND to describe the current state of the domestic industrial base for coal-based electricity generation. RAND was asked to focus on coal-fired EGU design and construction, technology development, and equipment manufacturing.

This monograph addresses the concern about whether the industrial base for the U.S. domestic coal-based electricity-generation industry can maintain the capability to design, construct, operate, and maintain coal-fired EGUs within reasonable cost, schedule, performance, environmental, and quality expectations. By first describing the capability that is inherent in the existing coal-fired fleet, this monograph takes a first step toward addressing the larger policy questions of how to develop, deploy, and maintain an advanced, low-carbon electricity-generation industry capability into the future.

We framed this research in terms of capability: What resources are required to sustain the capabilities of a financially and technically viable coal-based electricity-generation industry? Maintaining such a capability requires a combination of the following:

- enough firms that can design and manufacture the required unit components and subsystems, including those components unique to coal
- enough engineering, procurement, and construction (EPC) firms capable of designing and managing the construction of a coal-fired generating unit
- enough demand for services to enable EPC and component-manufacturing firms to remain financially viable
- enough skilled labor
- enough annual engineering, operation and maintenance (O&M), and construction market–sector activities to sustain the required level of experience in the skilled workforce.

To investigate these dimensions of the domestic industrial base, we reviewed existing reports and databases, conducted interviews with a variety of stakeholders, and collected and analyzed data describing key elements of industry capability and validation or verification of concerns.

Key Findings

Some Capabilities Are Maintained Through Active Operation and Maintenance and Pollution-Control Markets, but New Construction Is Required to Maintain Complete Capabilities

The coal-based electricity-generation equipment industry can be divided into three interconnected market sectors, distinguished by the type and scope of activities: new-unit construction, O&M, and pollution control.

The new-unit construction sector includes the design and construction of new coal EGUs, either on existing sites or on new sites, and the design, development, and production of the major subsystems and components of a coal-fired EGU (pulverizer, boiler, steam-turbine generators, pollution control, and cooling towers). After a coal-fired generating unit is constructed, the O&M sector performs maintenance activities that range from simple repair or routine maintenance of plant subsystems to replacement or refurbishment of major components or subsystems.

The existing coal-fired fleet contains a mix of units with and without advanced pollution-control equipment. The installation and upgrading of pollution controls in response to regulations creates the market for pollution-control equipment manufacturing and installation. The O&M and pollution-control markets appear to be fairly robust, while the new-construction sector activity is sporadic and relatively small. These O&M and pollution-control sectors utilize some, but not all, subsectors of the U.S. industrial base. For example, although manufacturing coal-unique equipment occurs largely outside the United States, engineering design and manufacturing of pollution-control equipment remains largely within the United States.

Globally, more than 800 gigawatts (GW) of total new coal capacity is forecast from 2015 to 2035, with 600 GW of this installed in China, a rate of 30 GW per year (EIA, 2010a). In contrast, EIA's *Annual Energy Outlook 2011* reference case forecasts only 2 GW of new U.S. coal-fired capacity additions from 2012 through 2035 (EIA, 2011a). The global marketplace still offers enough business for U.S. firms to exercise some critical capabilities. Yet, the dearth of new U.S. coal-based generating-unit construction results in a lack of continued development in the integrated design, specialized project management, scheduling, procurement, and labor skills associated with constructing a new conventional or advanced coal-fired generating unit.

Additionally, U.S. firms participating in predominantly foreign coal EGU construction will not fully develop the critical experience in cutting-edge combustion technology for advanced coal-based generation systems, such as ultrasupercritical, integrated gasification combined-cycle (IGCC) and oxy-combustion, which will be critical if the industry transitions to advanced coal EGUs with carbon capture and sequestration (CCS). Experience gained using local coals in China and other countries deploying these technologies will provide some experience, but additional expertise on the design implications of using various U.S. coals with these advanced combustion systems will be needed. These remain critical areas in research and development (R&D) for advanced coal-based power generation.

Coal Boilers Are Critical to the Coal Power Industrial Base, but the Equipment Market Is Global

Several components of coal-fired power generating units are utilized in other types of power plants or in other industries; the largest and most important coal-unique system is the boiler. It consists of components that are exposed to the ash and other combustion products of coal and need to contain steam at high pressures and temperatures. These components are called *pressure parts* or *pressure components*. They include seamless tubing, carbon-steel plates, forgings, and castings and are assembled into the economizer, interior furnace tubing, steam separators, headers, walls, superheaters, and reheaters that characterize the boiler system. Fabrication of pressure parts requires facilities capable of forging and applying heat-treatment techniques to the large components, as well as the use of advanced and highly skilled welding. Large components of the boiler system are fabricated in sections in factories and then assembled at the construction site.

The market for coal-based electricity-generation equipment and construction is global, and the United States no longer dominates in terms of demand for equipment or power plant construction. Most new construction activity occurs outside the United States, especially in China. Since 1990, coal power plant installation activity in China has consistently been more than 10 GW annually, with a peak of more than 80 GW in 2006 (EIA, 2010a). This compares with a high of 6 GW in 2010 in the United States, which is the highest annual demand in the past two decades. Our interviewees concurred that nearly all pressure-component manufacturing currently occurs outside of the United States, mostly in Asia, with some facilities in eastern Europe. However, several firms maintain a core engineering design and technology development capability within the United States. These firms include a few U.S.-headquartered firms that have manufacturing activities overseas.

The influence of overseas markets on both supply and demand is extraordinarily important to U.S.-based firms. This influence reflects shifts in market signals and changes in the business environment. Several reasons were given by interviewees for the shift to international pressure parts manufacturing, including the following:

- desire to colocate manufacturing supply with local coal-fired power plant demand
- reductions in labor costs and productivity gains achieved in international locations
- lack of available manufacturing facilities in the United States with steel heat-treatment capabilities suitable for very large coal power unit components
- a need for U.S. EPC firms to globally source major equipment in order to remain competitive in the U.S. market.

Some of these factors were also mentioned during our interviews on why many components of coal-powered EGUs besides pressure parts were also manufactured in Asia. Because delivery times to the construction site are established early in the project development process, sufficient time exists for custom fabrication and shipment from an international supplier. Thus, interviewees stated, barring large increases in logistics costs (or risks), market forces affecting domestic demand and manufacturing costs do not support major investments in domestic manufacturing capabilities to support new unit construction.

Workforce Challenges Are Not Unique to Coal but Would Raise Costs During High-Demand Periods

A 2009 industry survey found that roughly half the current workforce in several relevant job classifications will be eligible for retirement between 2009 and 2015, including 46 percent of pipefitters, pipelayers, and welders; 51 percent of technicians; and 51 percent of engineers

(Center for Energy Workforce Development [CEWD], 2009).[1] Perhaps more significantly, utilities responding to the survey reported having trouble filling openings in skilled-labor positions: "between 30 [and] 50 percent of applicants (those [who] met the minimum requirements for a position) were not able to pass the pre-employment aptitude test" (CEWD, 2009). Interviewees also supported this concern.

Although these results are not specific to coal, they do indicate that the utility industry perceives a current shortage of skilled labor in the workforce and anticipates that the shortage will grow worse in the near future. Bureau of Labor Statistics data, which show decreasing labor associated with fossil-fuel power plants, are consistent with the concern that the workforce is declining and new workers are not being hired at a one-for-one rate of replacement, but they do not explain the causes of this decrease in the workforce.

In interviews, individuals expressed concern that, should there be an increase in construction of coal EGUs, there would likely be shortages of skilled labor that would increase labor costs and slow the construction of coal-fired generating units. Of particular concern to interviewees were the supply, productivity, and costs of skilled welders critical to construction of coal-fired generating units. However, interviewees were confident that the industry could respond to correct labor shortages if a period of sustained demand were to arise. There are several reasons that they believed this to be the case:

- With a global marketplace for coal-fired generating units, the technical knowledge of knowing how to build a power plant will not be permanently lost with retirements.
- The markets for O&M and installation of pollution-control equipment, as well as construction of other electricity-generation technologies, such as natural gas–fired power plants, provide some level of relevant training for workers.
- If a sustained surge in demand occurred, firms might need to invest in training programs to increase the size and capabilities

[1] The survey covered 31 companies accounting for 44 percent of U.S. electricity and natural gas utilities. See CEWD, 2009.

of the workforce, and the supply of workers could be increased to meet a sustained demand within a few years.

In general, we found that the workforce responds to market forces in a way similar to the way it does in equipment manufacturing. Stakeholders share responsibility for the hiring and training of new workers, and the organization of existing workers, including utilities; trade unions; equipment manufacturers; architecture, engineering, and construction (AEC) and EPC firms; and community colleges and trade schools. In the face of a sustained increase in demand, according to their experience and interview discussions, these stakeholders are well-suited to collaborate to eventually generate the level of skilled labor needed. However, this adjustment phase is likely to be accompanied by increases in the cost of labor, albeit temporary, and a period of longer-than-normal construction times until efforts to increase the pool of skilled labor are implemented or demand subsides. One potential strategy for mitigating shortages of skilled labor is an increased emphasis on programs that increase awareness of career opportunities and provide training and apprenticeship opportunities, and several interviewees said that their organizations had active programs in these areas. There is potentially a role for federal support for these kinds of programs and partnership with the organizations that run them.

Interviewees also expressed concern that the near-term demand and short compliance timelines involved with installing equipment to meet proposed U.S. Environmental Protection Agency (EPA) pollution-control regulations will substantially increase construction costs and schedules for pollution-control retrofits and, in tandem, for new-unit construction. After the installation of pollution-control retrofits to existing units is complete, this market will be limited to pollution controls for new coal unit construction. Hence, interviewees noted that incentives for pollution-control firms to add capacity during the compliance period will be limited. This is an area that should continue to be examined for potential bottlenecks and strategies to alleviate workforce, cost, and schedule pressures.

Construction Costs and Schedules Have Increased for Recently Built Coal-Fired Power Plants

As with all aspects of construction, cost and schedule increases have been and will continue to be experienced during periods of high demand. Cost growth and schedule slip are common in large, complex projects (see Merrow, McDonnell, and Arguden, 1988; Merrow, 1989). Additionally, new coal-fired unit designs and resurgence of deployment will experience cost growth. If there is a resurgence of demand and it includes advanced coal technologies, then higher costs are expected until experience is gained with deploying these systems (Merrow, Phillips, and Myers, 1981). Many factors drive undesirable cost and schedule outcomes, from technical difficulty to poor management. In the limited coal-fired electricity-generating unit examples we were able to identify, increased cost and schedule are due at least in part to rising prices of materials and labor. In turn, these price increases are due in part to increasing global demand for materials and shortages of skilled labor required for construction. The IHS CERA Power Capital Costs Index and IHS CERA European Power Capital Costs Index show a near doubling of power plant costs in the past decade (IHS, undated). These data include coal, gas, and wind projects but exclude nuclear power. IHS stated that the slight rise in prices in 2010 was due to increased steel costs but was tempered by declines in major coal-equipment costs due to lack of orders and interest (IHS, 2010). EIA's updated capital cost report shows an increase in the capital costs for coal but not for natural gas, which, the report hypothesizes, could be due to the higher cost of financing for larger projects or to the lack of firms with experience constructing large megaprojects (EIA, 2010c).

Locally sourced labor for construction, engineering, and construction management services comprises about one-third of capital costs for coal power plants (Gerdes et al., 2010). In contrast, globally sourced procured equipment, which is highly sensitive to material input costs and market demand, represents nearly half of the capital costs for coal power plants. The cost pressures of the recent past are driven more by demand in China and elsewhere, which accounts for the vast majority of the market, than they are by increased U.S. demand. The role of potential shortages of skilled labor in the United States is more difficult to ascertain in terms of cost pressures. However, the decline in costs in 2010 and high unemployment in construction trades show that these pressures have largely eased.

Potential Future Challenges Could Increase Costs and Schedules

Some industry interviewees were concerned about the effect that a prolonged drought of a decade or more without new construction of coal-fired generating units would have on domestic labor (especially in project management and engineering) when coupled with large-scale workforce retirements. Others raised questions of how the effects of these trends, or even the short-term price oscillations associated with the current market, would affect the competitiveness of coal-fired power generation and thus have implications for the long-term costs of electricity generation in the United States.

From a policy perspective, the market for coal-fired generating units might be too narrow when examining potential problems. There is a robust market for electricity-generation technologies, and owners will choose the cheapest option (subject to regulations, fuel prices, local conditions, and labor costs). Most interviewees stated that the electricity sector has historically experienced large surges in power plant construction demand followed by periods of low activity and that, after a period of initial higher prices, the industry has generally adapted to the changing conditions of the market. It is unclear how quickly markets will adapt and whether prices will stabilize in future periods of high demand. Interviewees also noted that other electricity-generation technologies, such as natural gas, nuclear, and renewables, compete with coal for some of the labor, material, and equipment required for plant construction, and a large power-sector reinvestment that involves several technologies deployed simultaneously would initially result in cost and schedule pressures for these resources across all technologies.

If the industry transitions to advanced coal EGUs with CCS, additional cost and schedule challenges are anticipated, as with all first-of-a-kind technologies. We examined the state public utility commission testimony of one utility executive discussing the sources of cost growth for a specific recent IGCC power plant that is the first of its kind at this size (586 MW) (Haviland, 2011), which would facilitate the future inclusion of CCS. Unforeseen engineering challenges and changes, larger-than-expected requirements for material quantities, and delays increased costs for this project considerably. In his testimony, the official mentioned poor labor productivity as the project progressed, especially regarding specialized welding services. However, he attributes low labor productivity largely to the unanticipated quantities required and poor project and resource planning that resulted. Thus, workers often did not have the materials needed on-site for construction, which further affected productivity and costs. He also stated in his testimony that the unplanned increase in quantities substantially increased the number of welders needed and that the use of specialty alloys in the project caused welding difficulties (Haviland, 2011). Therefore, the project under consideration at the hearing at which he testified points to several potential causes for increased costs, but they are not clearly separable from first-of-a-kind-at-this-size learning issues. The industry should examine the experiences at this IGCC plant to improve designs, schedules, and workforce planning for future advanced coal power plants.

The preponderance of evidence examined for this monograph indicates that, despite the recent recession, the industry base to supply equipment for coal-based electricity generation remains capable of responding to potential demands in the near term. Ultimately, the interest in the capabilities of the coal-based power generation industry base is driven by questions of whether the industry will be able to respond to future demand. However, demand is highly uncertain and depends on forecasts of economic growth, coal power's competitiveness relative to other generation technologies, the success of technological development efforts to improve the efficiencies of and reduce the pollution associated with coal-fired and other power generation, and the path that future pollution and greenhouse gas regulations take.

Costs and project schedules are likely to increase when new coal power projects are initiated after a long period of low demand and during high-demand periods caused by short regulatory compliance timelines or market forces. Resource and workforce competition from constructing other electricity-generation technologies will likely contribute to these increases. Hence, this monograph is a first step in understanding the implications of a large-scale power-sector reinvestment and transition to advanced technologies for the U.S. industrial base. Although this and other studies have examined workforce and equipment needs for specific generation technologies, a holistic analysis, including multiple technologies deployed across a range of technology, demand, and construction scenarios, is needed to anticipate gaps in skills and resources and to determine robust strategies to alleviate barriers to a large power-sector reinvestment.

Acknowledgments

This research was funded by NETL. The contents of this work do not necessarily reflect the opinions of the research clients and sponsors. We thank NETL staff Kenneth C. Kern, Timothy J. Skone, Joel Theis, Joseph DiPietro, Peter C. Balash, and Charles Zelek for thoughtful comments. We are very grateful to the industry professionals who were interviewed for this work and who added their unique perspectives and insights. The monograph also benefited from helpful feedback from Keith Crane and David S. Ortiz of RAND and from Gregory F. Reed of the University of Pittsburgh. Finally, we thank the peer reviewers, Karen Obenshain of Edison Electric Institute and James T. Bartis, whose comments greatly improved the monograph. Any remaining errors are those of the authors.

Abbreviations

AEC	architecture, engineering, and construction
AEP	American Electric Power
AMP	American Municipal Power
BFB	bubbling fluidized bed
BLS	Bureau of Labor Statistics
CCS	carbon capture and sequestration
CEWD	Center for Energy Workforce Development
CFB	circulating fluidized bed
CO_2	carbon dioxide
DoD	U.S. Department of Defense
DOL	U.S. Department of Labor
DSI	direct sorbent injection
EEED	Environment, Energy, and Economic Development Program
EGU	electricity-generating unit
EIA	Energy Information Administration
EPA	U.S. Environmental Protection Agency
EPC	engineering, procurement, and construction
FGD	flue-gas desulfurization
GHG	greenhouse gas
GW	gigawatt

GWh	gigawatt-hour
HAP	hazardous air pollutant
Hg	mercury
HRSG	heat-recovery steam generator
IGCC	integrated gasification combined cycle
ISE	RAND Infrastructure, Safety, and Environment
kW	kilowatt
MACT	maximum-achievable control technology
MW	megawatt
NETL	National Energy Technology Laboratory
NO_x	nitrogen oxides
O&M	operation and maintenance
OECD	Organisation for Economic Co-operation and Development
OEM	original equipment manufacturer
PAC	powder-activated carbon
PC	pulverized coal
PSI	pounds per square inch
R&D	research and development
SCR	selective catalytic reduction
SNCR	selective noncatalytic reduction
SO_2	sulfur dioxide
SO_3	sulfur trioxide
TWh	terawatt-hour
USDA	U.S. Department of Agriculture

CHAPTER ONE

Introduction

Coal-fired power plants are the dominant source of electricity in the United States, providing 46 percent of U.S. electricity generation in 2010 (Energy Information Administration [EIA], 2011c). In light of the large share of coal-fired generating capacity in the power portfolio, much of the electricity consumed in the United States will continue to be generated from coal for an extended period of time. However, U.S. utilities have not invested heavily in new coal-fired generating capacity. Uncertainty about policy concerning reductions in emissions of greenhouse gases (GHGs) and conventional air pollutants has depressed investment in new coal-based electricity generation and upgrades to existing facilities. The GHG policy eventually adopted by the United States could change the costs of coal-based electricity generation through requirements for CCS or reuse of carbon dioxide or purchases of allowances or offsets and could make utilities reluctant to invest in coal-fired generating technologies. Emission standards for other air pollutants, as well as possible regulations addressing ash handling and water quality, would also affect the cost of electricity generation from coal. The international expansion of coal-fired power plants in China and other countries has shifted capital and labor assets offshore to support international demand. Additionally, electricity generation from other sources, including natural gas and renewables, has become more competitive in relation to coal. Finally, the deployment of new coal-fired electricity-generating units (EGUs), as well as maintaining the existing fleet, will require an industrial capacity to provide equipment, skilled labor, and project management expertise. But the workforce that provides and maintains the human and intellectual capital to support the industry is aging and shrinking. For these reasons, in the near term, the domestic industrial base to provide the capacity for future coal-based power units faces several challenges.

The market for construction of new coal EGUs is largely international. Most new construction activity occurs outside the United States, especially in China. For example, since 1990, coal EGU installation activity in China has consistently been more than 10 gigawatts (GW) annually, with a peak of more than 80 GW in 2006 (EIA, 2010a), in contrast to the 0–6 GW of annual installations in the United States. Globally, more than 800 GW of total new coal-fired capacity is forecast through 2035, with 600 GW of this to be installed in China, a rate of 30 GW per year. Growth of coal-fired electricity generation in China is projected to grow at more than 4 percent per year through 2035 (EIA, 2010a). In contrast, recent U.S. forecasts in the EIA *Annual Energy Outlook 2011* reference case show only 2 GW of new U.S. coal-fired capacity additions from 2012 through 2035 (EIA, 2011a). Such a gap in new construction of coal-fired power units in the United States would be similar to the ongoing hiatus in the construction of new nuclear-fired power plants.

The U.S. coal EGU equipment industry has adapted to the dearth of new orders for power units in the United States. However, if no new units are likely to be built in the foreseeable future, the U.S. industry is likely to have to make further adjustments. If, at some future point in time, U.S. utilities do decide to make substantial investments in new coal-fired capacity, will the U.S. industry be able to respond to such an increase in demand?

To better understand these challenges, the U.S. Department of Energy's National Energy Technology Laboratory (NETL) requested that RAND describe the current state of the domestic industrial base for coal-based electricity generation. RAND was asked to focus on coal EGU design and construction, technology development, and equipment manufacturing. By first describing the capability that is inherent in the existing coal-fired generation fleet, this monograph takes a first step toward addressing the larger policy questions of how to develop, deploy, and maintain an advanced low-carbon electricity-generation industry capability into the future.

Background

Several factors affect the structure and capabilities of the coal-based electricity-generation equipment industry, including the following:

- U.S. electricity demand, including baseload, load following, and peak demand requirements
- unit size, age, and expected useful life of the existing coal-fired power plant fleet
- price and cost of electricity, labor, fuels, and materials
- relative cost of producing electricity from different sources, especially natural gas
- investment requirements (new construction, operation and maintenance [O&M], and pollution-control upgrades)
- environmental regulations and the associated cost of compliance[1]
- commercialization and innovation of new coal-related technologies
- international technological trends and demand for equipment.

The demand for a coal-based electricity-generation capability is driven by both U.S. demand for electricity and the need to replace older capacity that is retired. However, the U.S. coal-based EGU construction industry is heavily influenced by the global market. The global market affects U.S. manufacturers' ability to maintain capabilities through external (or foreign) demand, international technological trends, and global competition for raw materials, workforce talent, and manufacturing, construction, and repair and refurbishment jobs.

Privately or publicly owned utilities are the primary owners and operators of coal EGUs. A utility's decisions about investments in coal-related market sectors are driven by the relative price of electricity and

[1] This includes GHGs, sulfur dioxide (SO_2), nitrogen oxides (NO_x), particulates, ash, ozone, water cooling, water quality, and mercury (Hg) and other hazardous air pollutants.

the cost of producing it from alternative sources of fuel, as well as the utility's ability to recover capital costs. Environmental regulations in the United States have a significant impact on the coal-based electricity-generation industry. Regulatory requirements to meet pollution-control standards over a broader set of potential waste streams can have a significant effect on the cost of building and operating a coal-fired power unit. The cost of complying with existing or anticipated regulations is often cited in utilities' decisions regarding upgrades and retrofits to existing units, and it affects decisions regarding coal or alternative electricity-generation technologies as the preferred source for future new capacity.

According to what we see in the decline in the number of announced new coal plants, U.S. utilities will add substantially fewer plants in the next five years than were previously planned (Shuster, 2011). Compliance costs associated with current and future regulations are likely to increase the cost of coal-based generation, making other generating technologies, especially natural gas, more attractive. Some of the technologies required to comply with potential GHG regulations, such as carbon capture and sequestration (CCS) and storage and reuse technologies, have not yet been commercially demonstrated at the scale required. Meanwhile, natural gas is readily available and often economically competitive at current prices. Yet, some U.S. utilities that own substantial amounts of existing coal-fired capacity and that are able to obtain coal at low cost have invested in or have plans for investing in new, large-scale coal-fired capacity, with about 7.6 GW under various stages of construction and 6.8 GW permitted in advance of construction, and additional coal power generating units in the early stages of project development (Shuster, 2011). Historically, however, most of the investments in new generating capacity in the United States in the past 20 years have been in natural gas and, to a lesser extent, renewables (Ventyx, 2011).

Suppliers of equipment and labor services are reacting to an increasingly competitive global marketplace for coal-based electricity-generation equipment. For more than two decades, demand for coal-based capacity has been increasing rapidly in a global market, an increase that is driven largely by demand in China. In contrast, the

U.S. domestic market for coal-fired power plants has historically gone through boom and bust cycles; the most recent boom was between 2004 and 2010, a period in which U.S. utilities brought online more than 14 GW of new coal-fired capacity (Ventyx, 2011). Construction is currently being completed on the remainder of units that were planned and designed during that time.

In addition to the challenges cited already, recent analyses have identified other factors that could negatively affect demand for coal-based electricity-generation equipment in the United States, including the following (Wood Mackenzie, 2010; Freese et al., 2011):

- electricity-generation reserve capacity, particularly in underutilized natural gas power plants
- concerns about obtaining favorable financing for large coal power plant projects.

These factors, largely external to the industry, affect coal-fired fleet investment and retirement decisions. The changing economics of coal-based electricity generation suggest that maintaining industry capability cannot be considered in a vacuum.

Approach

Can the U.S. coal-based electricity-generation equipment industry maintain the capability to respond to future U.S. demand for electricity by retaining the ability to design, construct, operate, and maintain coal-fired power units within reasonable cost, schedule, performance, environmental, and quality expectations? This study answered this question using a framework that describes the resources and activities that are required to sustain a financially and technically viable coal-based electricity-generation capability in the United States. Maintaining such a capability requires a combination of the following:

- enough firms that can design and manufacture the required EGU components and subsystems, including those components unique to coal
- enough engineering, procurement, and construction (EPC) firms capable of designing and managing the construction of a coal EGU
- enough demand for services to enable EPC and component-manufacturing firms to remain financially viable
- enough skilled labor
- enough annual engineering, O&M, and construction market–sector activities to sustain the required level of experience in the skilled workforce.

To investigate the dimensions of the domestic industrial base for this monograph, we did the following:

- reviewed existing reports and databases that describe elements of the coal-based electricity-generation industry's capabilities and began to identify potential challenges to the industry
- conducted interviews with a variety of industry- and workforce-related organizations to identify the key aspects of the industry (or capability), identify coal-unique capabilities, and further refine the challenges and possible problems that the industry faces
- collected and analyzed information describing key elements of industry capability and validation or verification of the concerns.

The coal-based electricity-generation equipment industry operates within the broader economic context of electricity generation in general. The industrial base for coal-fired electricity includes the equipment manufacturers that provide EGU components; the architecture, engineering, and construction (AEC) and EPC firms that design and manage generating-unit construction; and the skilled, unskilled, and professional labor that is employed by these companies to maintain the existing fleet, refurbish or upgrade units, and construct new capacity.[2] We have excluded labor and equipment for processes that are not unique to coal-fired electricity because competitive markets for these sectors are not dependent solely on activity in coal-fired electricity. These include the mining, processing, and transportation of coal, and power plant equipment and labor that are ubiquitous across power plant types.

Although there are no publicly available comprehensive reports addressing the full range of topics inherent in describing coal-based electricity-generation industry capabilities, some organizations have produced reports on specific aspects of the industry. Topics include workforce-related issues, financial or economic considerations, and technical aspects of current and future technologies associated with coal use for power generation. We also obtained both qualitative and quantitative data from government, industry, and stakeholder organizations, which allowed us to describe the existing coal EGU fleet and identify the key domestic, foreign, and U.S. subsidiaries of foreign companies that comprise the industrial base for coal-fired electricity generation in the United States. We examined the historical patterns of mergers, acquisitions, and new starts that characterize the evolution of this industrial base. We also assessed workforce trends at an aggregate level and the implications for knowledge, skills, and experience.

The information from existing reports and databases allowed us to develop a preliminary list of possible challenges to maintaining a financially and technically viable coal-fired electricity-generation industrial capability and to identify the key stakeholders we wanted to interview in order to refine this list of potential challenges. We conducted 25 structured interviews with a cross-section of industry stakeholders to complement and verify the quantitative data assessment (see Table 1.1).[3] Interviewees included major utilities with consider-

---

[2] We use *AEC/EPC* in this monograph to denote the capabilities of this coupled sector. Many AEC firms also serve in EPC roles, depending on contract structures chosen by coal EGU owners.

[3] We contacted many more organizations than actually agreed to participate in the study. All interviews were conducted as confidential, not-for-attribution discussions.

Table 1.1
Number and Types of Organizations Interviewed

Category	Interviews
Equipment manufacturers	6
Workforce-related organizations	4
AECs/EPCs	6
Utilities	3
Industry and professional organizations	6

able coal-fired power plant assets, leading AEC/EPC firms constructing coal-fired power units, original equipment manufacturers (OEMs) providing major components and subcomponents for coal-fired electricity units, critical-material firms supplying specialty materials to the power industry, organized labor associations, industry trade and research organizations, and engineering and workforce associations.

Interviews with stakeholders focused on obtaining their views on the following:

- the main concerns and problems that the coal-based electricity-generation industrial base is facing
- the primary suppliers of products and services for coal-powered generation equipment, both foreign and domestic
- the impact and implications of the global market for coal-based electricity-generation equipment
- project management–related issues, including supply chain, product quality, cost, and timing
- the sources and potential effects of current and anticipated regulations affecting coal EGUs
- workforce-related concerns and problems, including demographics, skilled-labor availability and overall supply, hiring, training, skill sets, and experience levels
- the relative uniqueness of technologies, systems, subsystems, and skilled labor to coal-fired electricity generation.

These topics were used to guide the discussions. Some stakeholders had more interest, and expertise, in specific areas.

We then attempted to verify or validate the issues and concerns raised in the interviews with more-objective empirical evidence. The availability of such information constrained our ability to definitively support some concerns; however, the existence of common themes among independent stakeholders, existing reports and analysis from third-party organizations, and available databases provided confidence in our findings and observations.

Outline of This Monograph

This monograph is structured as follows. Chapter Two presents a description of the existing U.S. coal EGU fleet, including coal's contribution to U.S. electricity generation and specific characteristics of the fleet, such as age and installed pollution controls. Chapter Three discusses the major components and subsystems of coal-fired power units and identifies components that are common with other electricity-generation technologies, as well as those that are unique to coal. In Chapter Four, we discuss the structure of the coal-based electricity-generation industrial base and the primary market sectors it supports. Additional information on firms in the domestic coal-fired electricity industrial base is included in Appendix A. Chapter Five presents available information on relevant workforce trends and analysis, as well as a discussion of training and other workforce-related issues. Chapter Six summarizes our conclusions and observations and identifies questions regarding coal-based electricity-generation industry capabilities that remain unanswered. Finally, Appendix B provides supplemental data on the electricity-sector industrial base's workforce.

CHAPTER TWO

The Role of Coal-Fired Power in the U.S. Electricity Sector

Coal is a major source of energy for the United States, providing more than 21 percent of the total energy consumed in 2010 (EIA, 2011c). Since the middle of the 20th century, annual U.S. coal consumption has doubled from half a billion tons largely consumed by industry, to more than 1 billion tons largely consumed by electric power generation in 2010, as shown in Figure 2.1. Increasing demand for electricity, coupled with coal's high energy density and ease of shipment by rail and

Figure 2.1
Annual U.S. Coal Consumption, by Sector, 1949–2010

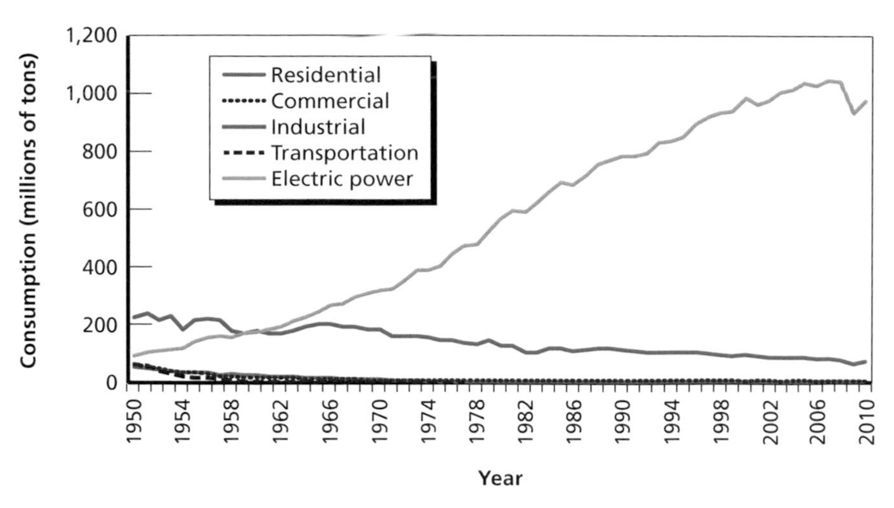

SOURCE: EIA, 2010b, 2011c.
RAND *MG1147-2.1*

barge, facilitated the deployment of coal power plants in the 20th century. During the same period, industrial energy use continued to shift toward natural gas or to offshore locations. In 2010, 93 percent of U.S. coal consumed was used in the electric power sector (EIA, 2010b).

Electricity Supplied, by Fuel Type

The 338 GW of grid-connected coal-fired power plants represent 33 percent of the approximately 1,000 GW of installed capacity of U.S. power plants (Ventyx, 2011; EIA, 2010b). The remainder of the electricity fleet capacity is comprised of natural gas (39 percent), nuclear (10 percent), hydropower (8 percent), petroleum (6 percent), and non-hydro renewables (4 percent). In characterizing the industrial base for coal-fired powered electricity, it is important to distinguish the power plant fleet's installed capacity, represented in gigawatts, from the actual annual electricity generation provided by each fuel, represented in gigawatt-hours (GWh). Because of the higher utilization of coal-fired power units in 2009, with only one-third of the installed capacity, coal-fired electricity represented about 45 percent of the gigawatt-hours generated in the United States in 2009, while natural gas represented about 23 percent of generation (EIA, 2010b).[1]

The excess capacity designed into the electricity sector, coupled with competition between fuels for generation across hourly, daily, and seasonal schedules, also has implications for the coal-fired power-equipment industrial base. Depending on fuel prices and other characteristics, electricity generated by natural gas could displace higher-cost coal. Since 1990, coal-fired generation has declined from more than 52 percent of generation to 45 percent in 2009, while natural gas generation has nearly doubled from 12 to 23 percent of total generation, as shown in Figure 2.2. To meet new demand for electricity in

---

[1] The rate of utilization relative to a power plant's installed capacity is called its *capacity factor*. Like baseload coal power plants, nuclear power plants also typically have high capacity factors. In contrast, power plants fired by fuel oil and other refined-petroleum products are used sparingly, accounting for only about 1 percent of total generation in 2009 (EIA, 2010b).

Figure 2.2
Electricity Net Generation, by Fuel Type

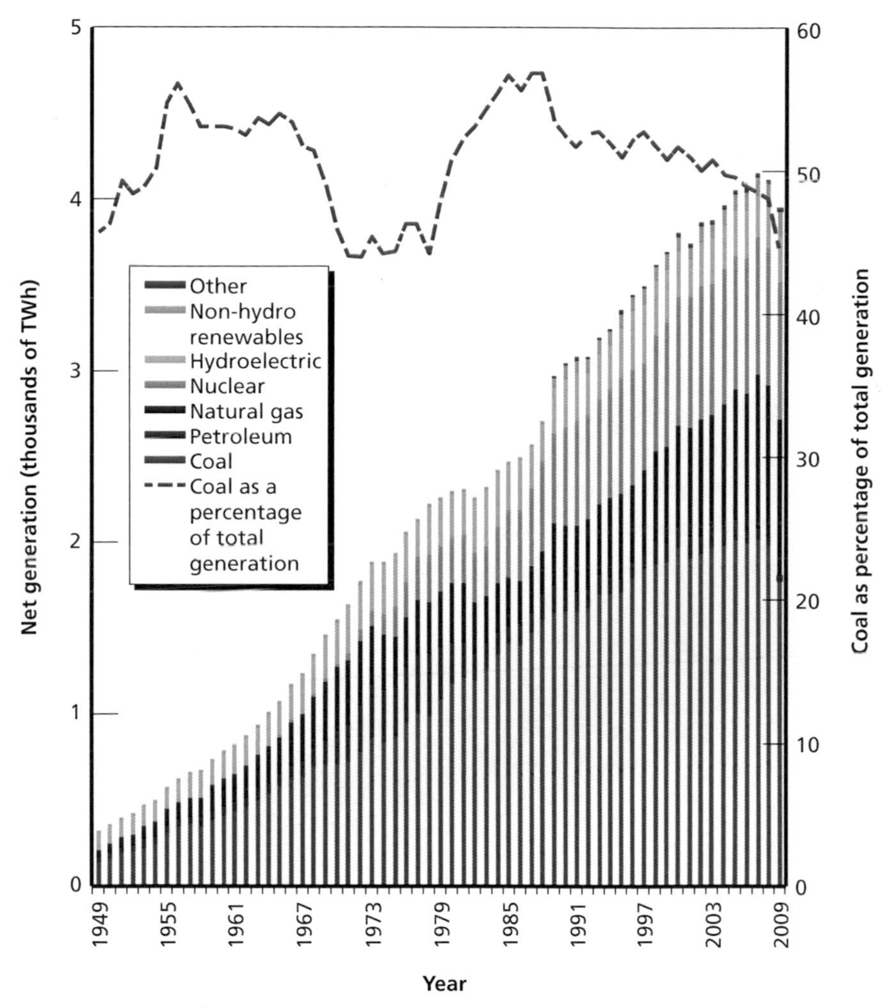

SOURCE: EIA, 2010b.
NOTE: Coal-electric power-sector net generation increased to 46 percent of total generation in 2010 (EIA, 2011c). TWh = terawatt-hour.
RAND *MG1147-2.2*

the short term, utilities evaluate the opportunities and risks associated with increasing output from existing units or implementing demand response and efficiency measures with customers. In the long term, to meet demand increases not abated by efficiency measures, utilities plan for adding new capacity and assess suitable power plant technologies.

Description of the Coal-Fired Power Plant Fleet

Capacity and Age of Existing Units

The U.S. coal-fired power plant fleet consists of 1,158 generating units representing about 338,000 megawatts (MW) of capacity (Ventyx, 2011). A coal EGU, which generally consists of coal and water feed systems, boiler, steam-turbine generator, pollution controls, and other systems, can vary in nameplate capacity from less than 50 MW to about 1,300 MW, and a power plant can contain more than one unit. Figure 2.3 shows the years in which currently operating coal-fired power capacity came online.[2] Coal was already the dominant fuel for electricity generation in the middle of the 20th century, but the coal-

Figure 2.3
Currently Operating Coal-Fired Power Plant Capacity Additions Over Time

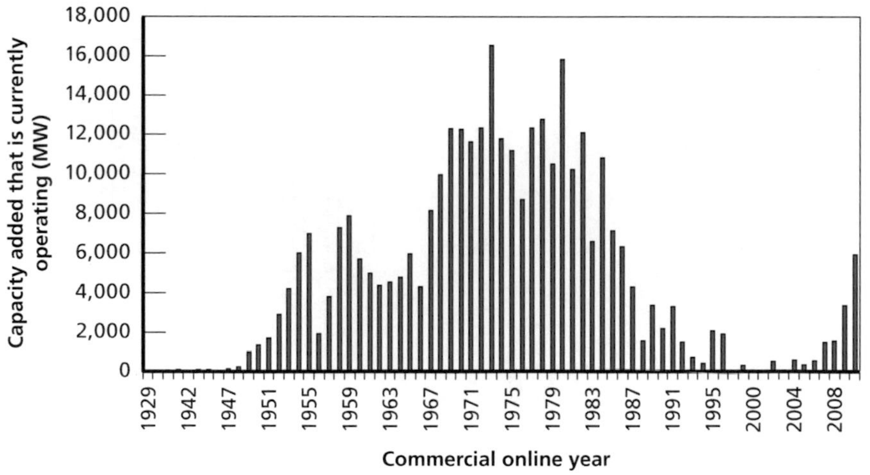

SOURCE: Ventyx, 2011.
RAND *MG1147-2.3*

[2] *Nameplate capacity* refers to the maximum rated power output of an EGU as designed by the manufacturer.

fired power fleet experienced a large, sustained construction boom from 1965 to 1985, with two-thirds of the currently operating coal-fired power units coming online during that period. As a result, most of the coal-fired generating units are 25 to 45 years old. Major components of coal-fired generating units require repair and refurbishment, depending on age and use, so older units would generally have higher O&M requirements and can lack many of the advanced control systems that reduce multiple pollutants to levels achieved by more-recent units.

Relatively few coal-fired generating units were brought online between 1990 and the present. In some years, no new units were brought online; in other years, several. The approximately 6,000 MW that came online in 2010 includes 11 separate units (Ventyx, 2011). In the past two decades, utilities have chosen to operate older coal-fired units longer than originally expected and have built new natural gas–fired units largely to satisfy new demands for capacity and replace coal-fired capacity that has been retired (Ventyx, 2011). Nevertheless, because of the large base of coal-based electricity-generation capacity, coal EGUs will continue to generate a substantial share of U.S. power for the foreseeable future, assuming that electricity demand is either flat or increasing.

The capacity of coal EGUs in the fleet also affects the coal-fired equipment industrial base. Installation of new units, regardless of their size, requires design and construction services and equipment. Larger-capacity units would increase the amounts of equipment and material needed for each project, the skilled labor required, and, generally, the total project duration. Yet, the total number of design and construction personnel would likely not scale linearly with unit size. As shown in Figure 2.4, the majority of existing units in the coal-fired fleet are small. Further, of the more than 475 units smaller than 150 MW, about 200 units were smaller than 50 MW. Although newly installed units tend to be larger, the market still demands some small units, as discussed in Chapter Four.

Figure 2.4
Sizes of Coal-Fired Electricity-Generating Units in the Existing Fleet

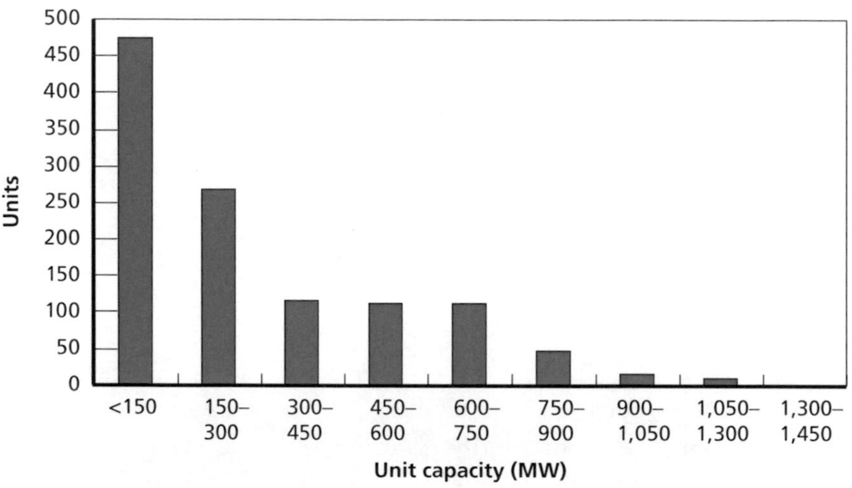

SOURCE: Ventyx, 2011.
RAND *MG1147-2.4*

Pollution Controls Installed on Existing Coal-Fired Units

Regulations require reductions of pollution generated from coal-fired electricity, primarily SO_2 and NO_x, as well as particulates, mercury, and other pollutants. Depending on the age, size, and location of a coal EGU, an existing unit might not have the required level of controls. The age of the coal-fired fleet and the proportion of units without one or more of the currently required pollution-control technologies largely define the market for retrofitting existing units with pollution-control systems. As discussed in Chapter Four, some of the same firms and workforce assets are utilized in retrofitting existing units with pollution controls. However, many of the critical design, manufacturing, and construction activities in the coal-fired industrial base are exercised only in the construction of new units. Figure 2.5 shows the proportion of coal-fired units with and without significant SO_2-reduction controls. Nearly 60 percent of existing units (or more than 40 percent of the existing capacity) do not have SO_2 controls, and these are typically the oldest units. Recent and proposed updates to pollution-control regulations will increase the number of pollution-control retrofits under-

Figure 2.5
Number of Currently Operating Coal Electricity-Generating Units with and Without Sulfur-Dioxide Controls

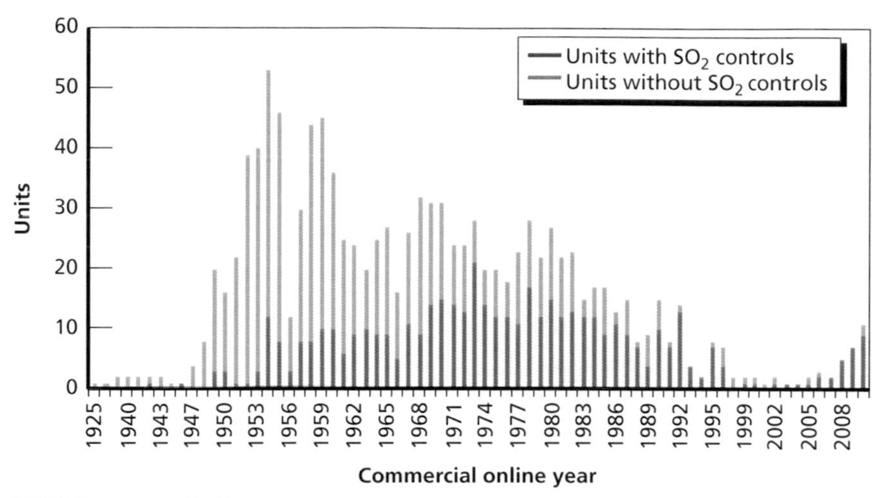

SOURCE: Ventyx, 2011.
RAND *MG1147-2.5*

taken, but, ultimately, the retrofit market will diminish as upgrades are completed.

Growth in Capital Costs of Recently Built Coal-Fired Units

The capital costs of installed coal power plants, which include equipment procurement, materials, engineering and construction management services, and skilled construction labor, have increased considerably in the past decade. Cost growth is common in large, complex projects (see Merrow, McDonnell, and Arguden, 1988; Merrow, 1989). Many factors drive cost growth, from technical difficulty to poor management. Figure 2.6 shows the IHS CERA Power Capital Costs Index (which covers North America) and IHS CERA European Power Capital Costs Index, which track the costs for materials, equipment, design, and construction of power plants. The indexes show a near doubling of power plant costs in the past decade in both North America and Europe (IHS, undated). These data include coal, gas, and wind projects but exclude nuclear power. The increases in power plant costs in this

Figure 2.6
North American and European Power Plant Capital Cost Indexes

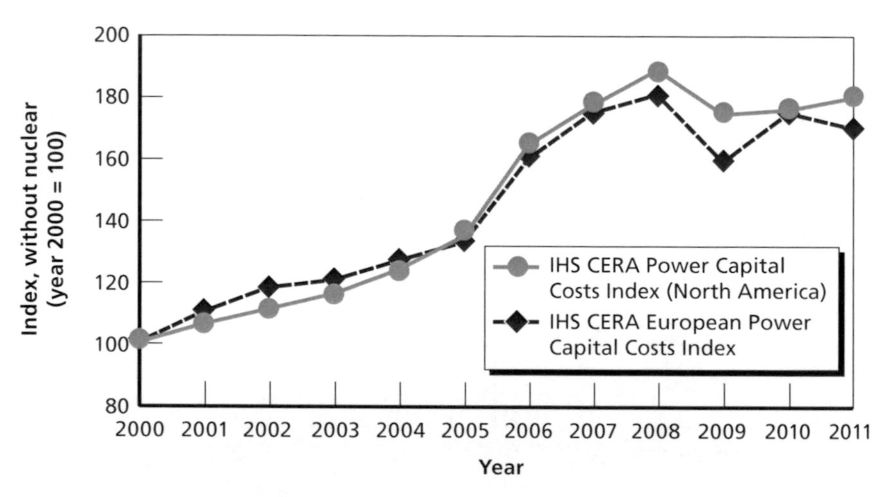

SOURCE: IHS, undated.
RAND *MG1147-2.6*

period mirror broader increases in construction services and materials due to global demand during this period. IHS stated that the slight rise in prices in 2010 was due to increased steel costs but was offset by declines in coal-fired major equipment costs due to lack of orders and interest (IHS, 2010). It notes that price increases in late 2010 and early 2011 were driven by demand for equipment, material, and labor in industrial activity outside the power sector. These increases were tempered by decreases in costs for power-plant major equipment, which fell by 1 percent in early 2011 (IHS, 2011). Although the recent slight decline in European costs was largely the result of exchange-rate fluctuations, IHS CERA stated that power-plant construction-labor costs for Europe and North America were tracking one another (IHS, 2011).

EIA estimates for the cost of deploying coal EGUs have increased substantially, as Table 2.1 shows. EIA's updated capital-cost report shows an increase in the capital costs of coal power plants. EIA's estimates for natural gas plants have not exhibited the same cost increases, which, the report hypothesizes, could be due to a higher cost of financ-

Table 2.1
U.S. Department of Energy Estimates for Coal Power Plant Capital Costs

EIA Annual Energy Outlook Year	2010 $/kW
2007	1,327
2008	1,635
2009	2,145
2010	2,257
2011	3,167

SOURCE: Freese et al., 2011.
NOTE: kW = kilowatt.

ing for larger coal projects or to a lack of firms with experience constructing coal power plant megaprojects (EIA, 2010c).

We searched for recent U.S. examples of coal power plant cost growth to verify these trends and to potentially understand the factors contributing to the rise in prices. Publicly available information at the project level is relatively scarce, but there is some anecdotal evidence from existing reports, the trade literature, and local news accounts. For instance, one report attributes increases in construction costs for coal EGUs to "a significant increase in the worldwide demand for power plant design and construction resources, commodities and equipment" (Schlissel, Smith, and Wilson, 2008, p. 6). This was confirmed both by our interviewees and by the observation of price increases experienced in the United States and Europe (see Figure 2.6). The increase in demand is attributed to demand in China and India. Other factors driving cost increases mentioned in the report include an increase in U.S. demand for power plant and pollution-control equipment, resource competition from petroleum refining and nuclear power industries, and the "limited capacity of EPC firms and equipment manufacturers." The same report includes the following examples:

- Duke Energy's Cliffside Project was originally expected to cost $2 billion in 2006 for a two-unit plant. Later in 2006, Duke announced a 47-percent cost increase for the two-unit plant. As of 2008, the project was expected to cost $1.8 billion for the one unit that Duke will actually build (Schlissel, Smith, and Wilson, 2008, p. 1).

- The cost of a 960-MW plant for American Municipal Power (AMP) Ohio increased from just over $1 billion in October 2005 to just under $3 billion in January 2008 (Schlissel, Smith, and Wilson, 2008, Figure 2, p. 3). Construction of this plant was eventually canceled in 2009, when the estimated price of electricity would have been $4,125 per kilowatt in 2010 dollars (Freese et al., 2011, Figure 11, p. 35).
- The cost of a 300-MW circulating fluidized bed (CFB) plant for Wisconsin Power and Light increased 40 percent.
- The cost of Duke Energy's Edwardsport integrated gasification combined-cycle (IGCC) plant increased 18 percent between the spring of 2007 and April 2008. The cost increase was attributed to "unprecedented global competition for commodities, engineered equipment and materials, and increasing labor costs" (Schlissel, Smith, and Wilson, 2008, p. 3).
- The cost of Kansas City Power and Light's Iatan 2 unit increased 15 percent (Schlissel, Smith, and Wilson, 2008, p. 4).
- In 2007, Tenaska Energy canceled a coal-fired power plant project due partly to increased prices for steel and other construction costs (Schlissel, Smith, and Wilson, 2008, p. 5).
- In December 2006, Westar Energy deferred construction of a 600-MW plant due in part to increases in equipment costs. Westar attributed the equipment price increase to the fact that equipment manufacturers were producing at capacity and had no plans for expansion (Schlissel, Smith, and Wilson, 2008, p. 5).

The economic environment of the recent past has also deterred some EPCs from agreeing to fixed-price contracts for the construction of coal EGUs due largely to the risk of increasing prices for materials and "a tight labor market" (Schlissel, Smith, and Wilson, 2008, p. 4). Our interviewees confirmed that long-lead items, such as boilers and boiler components and steam-turbine generators, experienced both price increases and schedule delays during the recent surge in power plant investment.

What is important here is that, in these coal-fired power plant examples, cost growth is due at least in part to rising prices of materials and labor and that these price increases are in turn due in part to increasing global demand for materials and construction. It is challenging to determine the dominant factors, as well as to understand price increases that are long term and structural versus those that will abate in the near term with time and market signals.

CHAPTER THREE

Coal-Fired Power Plant Designs, Systems, and Components

Conventional coal EGUs have various designs and configurations but have similar processes. First, units receive, process, and combust coal to produce steam. This steam drives a turbine generator to produce electricity that is fed into the electricity transmission system. Cooling systems condense the steam back into water for reuse in creating steam. Finally, pollution-control systems reduce the levels of conventional air pollutants in the exhaust gases that are released from the facility's smokestack, and handling systems dispose of ash and other wastes. Units can be designed as subcritical, supercritical, or ultrasupercritical units, with supercritical and ultrasupercritical units operating at higher steam temperatures and pressures and yielding higher efficiencies (Kitto and Stultz, 2005; American Electric Power [AEP], 2009).

There are three primary designs for utilizing and combusting coal in EGUs. The current dominant design is a pulverized-coal (PC) unit, in which coal is ground via a pulverizer to a fine powder and then combusted in a burner.[1] PC units are the most prevalent technology in the coal EGU fleet and represent about two-thirds of the current units progressing and under construction (Shuster, 2011). An alternative design is a fluidized bed, in which coal is combusted above a bed of limestone or other material mixed by forced air, providing reductions in SO_2 and NO_x emissions. Fluidized bed power units can be designed as either a CFB used primarily for large coal-fired power units or a bubbling fluidized bed (BFB) used primarily for waste coal or biomass

[1] Cyclone boilers are similar to PC boilers but do not require extensive fuel grinding and are used for coals, wastes, and other fuels not suitable for PC operations.

(Kitto and Stultz, 2005). Combined, they represent a smaller portion of the coal-fired electric power fleet and about 20 percent of the current units progressing and under construction (Shuster, 2011). Coal gasification, a technique in which coal is processed in an environment that controls oxygen, temperature, and pressure, is another pathway for coal to be used for electricity generation. This procedure generates gas by-products that are combusted in a gas turbine, with the hot exhaust used to generate additional power in a heat-recovery steam generator (HRSG), as part of an IGCC power plant. IGCC designs provide opportunities to capture and sequester CO2 prior to the combustion stage, with the installation of additional equipment (Gerdes et al., 2010). The viability of IGCC plants has been demonstrated (see Tampa Electric Company, 2002) but remain an emerging technology in commercial applications (see Duke Energy, undated). Hence, IGCC plants do not represent an appreciable portion of the current domestic industrial base for coal-fired electricity as outlined in this monograph but could become important going forward.

Figure 3.1 depicts the process of generating electricity from coal-fired power plants, and Figure 3.2 lists the distribution of costs of major equipment systems for conventional coal EGUs.[2] These systems are comprised of a series of components, which can differ slightly depending on unit design and technology. Because PC units and, to a lesser extent, CFB units are the dominant designs for the current domestic industrial base, we focus on these units. Figure 3.2 also shows an approximate distribution of total equipment costs in a conventional supercritical PC plant, with the boiler and pollution-control systems comprising more than half of total equipment costs. Equipment costs are only one component of total plant costs, which are discussed in Chapter Four.

The most-important components in our analysis are those specific to coal-fired electricity generation. We posit that components used across a range of industries are much less vulnerable to an extended period of low levels of coal generating unit construction and therefore

[2] Detailed lists of the system components in reference PC coal plants are presented in Gerdes et al., 2010.

Figure 3.1
Process of Generating Electricity from Pulverized Coal–Fired Power Plants

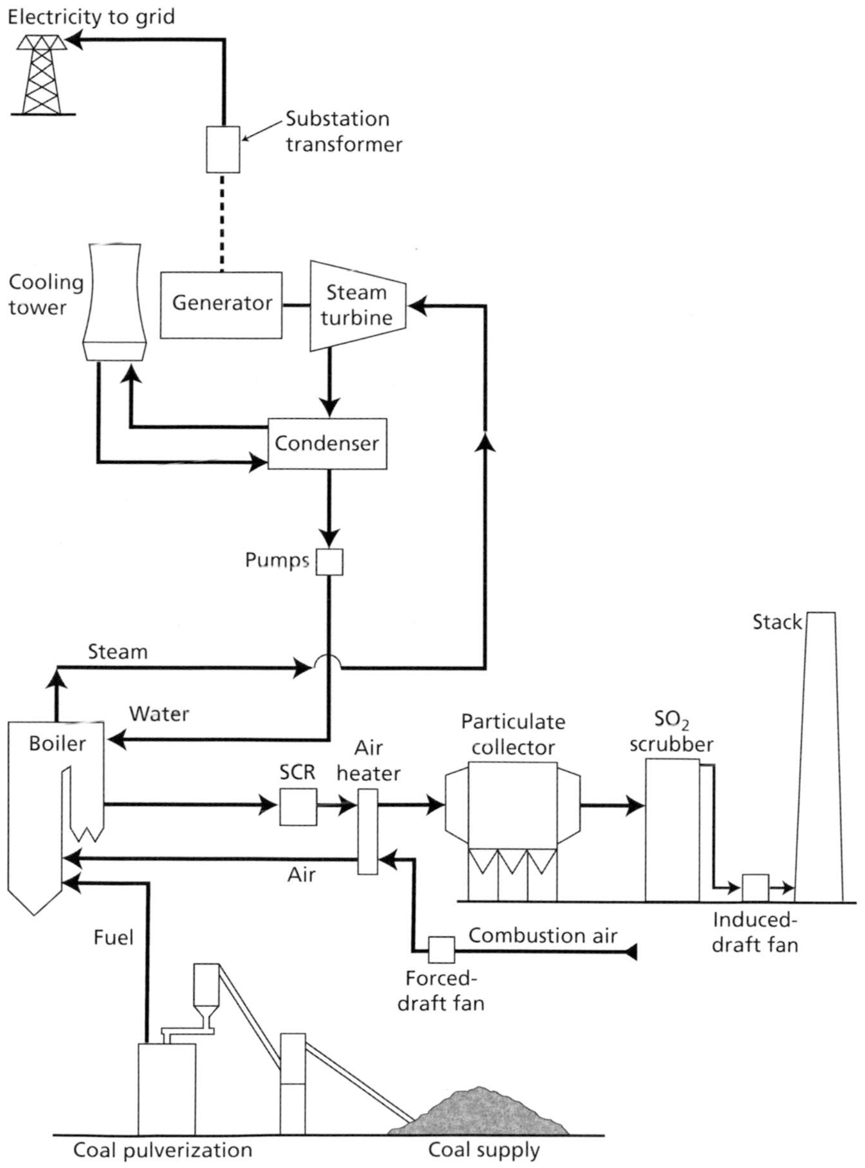

SOURCE: Adapted from Kitto and Stultz, 2005.
NOTE: SCR = selective catalytic reduction.
RAND *MG1147-3.1*

Figure 3.2
Distribution of Total Equipment Costs in a Conventional Supercritical Pulverized Coal–Fired Power Plant

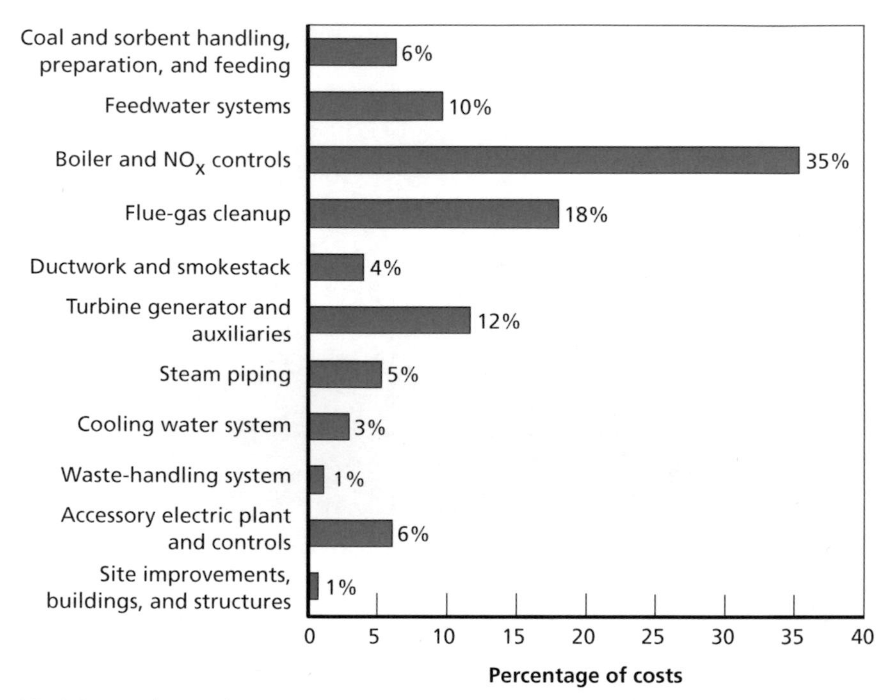

SOURCE: Gerdes et al., 2010.
RAND *MG1147-3.2*

are not of concern within the scope of this study. Our interviewees confirmed this conjecture. Several components of coal EGUs are utilized in other types of power plants or in other industries. Components that do not use specialty metals to withstand certain high steam temperatures and pressures or those that do not involve large customized pieces are also typically utilized in other applications. These common components are sold in larger, more diverse markets; supply chains and participants would be less affected by significant fluctuations in the market for coal-fired EGUs. Coal-handling and other material equipment, including conveyors and rock crushers, is used in other industries besides coal-fired electric power. Coal pulverizers are the notable material-handling component largely unique to coal. Elements of the

Rankine steam cycle, such as steam-turbine generators and feedwater and cooling systems used in conventional coal-fired power plants to produce power, are also used in other thermal steam power systems, such as natural gas combined-cycle operations. Although temperatures and pressures might be different in these other thermal power generation systems, our interviewees confirmed that active firms in steam-turbine generator and cooling-system markets possess capabilities and knowledge across generation fuels. Instrumentation, controls, accessory electrical equipment, and other systems are also utilized in other types of EGUs and not specific to coal power. Descriptions of major coal-unique components are presented in the next section.

Major Coal-Unique Components

Coal Pulverizers

For PC units, coal is fed into a pulverizer to be ground into a fine powder to facilitate combustion. Pulverizers are approximately 30 feet high and are fabricated mostly from steel and steel components. Several independent pulverizers are typically installed to accommodate firing and reliability requirements for an EGU. Pulverizers use metal crushers, separate compartments, and forced air to mechanically grind the coal to the desired fineness, ensure that larger particles are reground, and remove moisture from the coal before firing (Kitto and Stultz, 2005). One firm that manufactures coal pulverizers operates its own foundry to ensure the casting quality of high-wear pulverizer equipment and uses proprietary metal blends to improve component strength (Kitto and Stultz, 2005). Rock crushers and pulverizers are utilized in other industries, such as ore processing, cement manufacturing, and chemical manufacturing, and they could potentially be classified as a common component with a broad supply chain. However, because of the detailed specification requirements, specialty metals used, and critical role of pulverizers in coal-fired electricity generation, we list coal pulverizers as a coal-unique component.

Coal Burners

For PC units, after the coal has been pulverized, it is blown through a burner for combustion. The burner controls the flow of coal and air and ignites the mixture as it enters the boiler. There are typically four to eight burners for each pulverizer, with at least four pulverizers serving a large boiler. By reducing the flame temperature and the amount of oxygen during the early portion of coal combustion, considerable reduction in the formation of smog-forming NO_x can be achieved (Kitto and Stultz, 2005). Specialized burners that create these conditions are called *low-NO_x burners*, and these are the primary burner types installed in new PC units. Additionally, nearly all burners in existing coal-fired power plants have been upgraded to low-NO_x burners. Controlling NO_x formation during combustion is a critical design capability for burners and boilers.

Coal Boilers

Boilers burn a variety of fuels in many industries, but boilers for coal-fired electricity are large, customized integrated systems constructed with specialty metals and equipment. The coal boiler system for electric power units is the most critical coal-unique component in the industrial base, representing approximately 35 percent of a coal EGU's equipment costs (Gerdes et al., 2010). Large components of the boiler system are fabricated in sections in factories and then assembled at the construction site. The materials used are primarily steel, specialty materials that withstand high steam temperatures and pressures, and other metals. Some modular boiler components weigh 400–1,000 metric tons. A power plant boiler, heat-transfer equipment, and auxiliary components can weigh nearly 11,000 metric tons, requiring the equivalent of 500 railroad cars to deliver the materials to the construction site (Kitto and Stultz, 2005).

Current conventional coal EGUs can be designed as either subcritical or supercritical units. Nearly 75 percent of existing coal-fired EGUs in the United States are subcritical units, which operate at steam pressures of about 2,400 pounds per square inch (psi) and temperatures of about 1,050°F. The remaining 25 percent of the U.S. fleet consists of supercritical units, which operate at higher steam pressures and temperatures (3,500 psi and 1,050–1,100°F), increasing the efficiency and, generally, the capital cost of the unit (Kitto and Stultz, 2005; Gerdes et al., 2010). An emerging technology with higher efficiency is the ultra-supercritical unit, which operates above 3,500 psi and 1,100°F (AEP, 2009). The primary differences between supercritical and subcritical plants are (1) the specialized materials in the pressure parts designed to withstand the higher steam temperatures and pressures of supercritical operations and (2) the circulation systems used to move steam or water through the boiler system. Although no ultrasupercritical coal units exist in the United States, China opened its first ultrasupercritical coal unit in 2006 and has since deployed several others (Sun, 2010).

The coal boiler system is primarily comprised of low–steam temperature pressure components, the furnace enclosure, the high–steam temperature pressure components, air heaters, fans, and a section to reduce NO_x emissions, which is typically an SCR or selective non-catalytic reduction (SNCR) system (Kitto and Stultz, 2005). The low-temperature pressure components are a series of tubing and vessels that use the hot gases of the boiler to heat water as it enters and travels through the interior of the boiler. These low-temperature components include economizers, interior furnace tubing, steam separators, and headers and are composed of seamless tubing, carbon-steel plates, forgings, and castings. The furnace enclosure is the primary housing of the boiler, with the exterior composed of structural steel, sheet metal, and insulation, while the coal burners combust coal into hot gases into the interior of the furnace enclosure. Similar to the low-temperature pressure components, high-temperature pressure components are typically seamless tubing, carbon-steel plates, forgings, and castings. However, higher–carbon steel alloys are used to accommodate the higher temperatures. The high-temperature pressure components include superheaters and reheaters to maximize the temperature conditions of the steam cycle for power generation. Air heaters are a series of tubes or steel sections that use the hot flue gases to heat the air that is being fed into the boiler for combustion, thereby increasing efficiency. The SCR is part of the pollution-control system, but, because it is located within the boiler exhaust, it is categorized as within the boiler system. The SCR or SNCR system injects ammonia into the flue gas to remove

NO_x. Because both of these NO_x-control systems are used in other types of power plants, the technologies are not unique to coal, but their size, application, and integration for new and retrofitted boilers require specialized professional services. Figure 3.3 depicts a schematic of a supercritical PC boiler, highlighting some of the major components discussed in this section.

Boilers are designed for specific coal types and can encounter slagging, fouling, ash-handling, or other problems if a different coal type is used (Kitto and Stultz, 2005). Depending on coal and combustion characteristics, ash can deposit in boilers, erode or corrode heating surfaces, and pose difficulties for removal. Hence, avoiding ash chal-

Figure 3.3
Schematic of Supercritical Pulverized-Coal Boiler for Electricity Generation

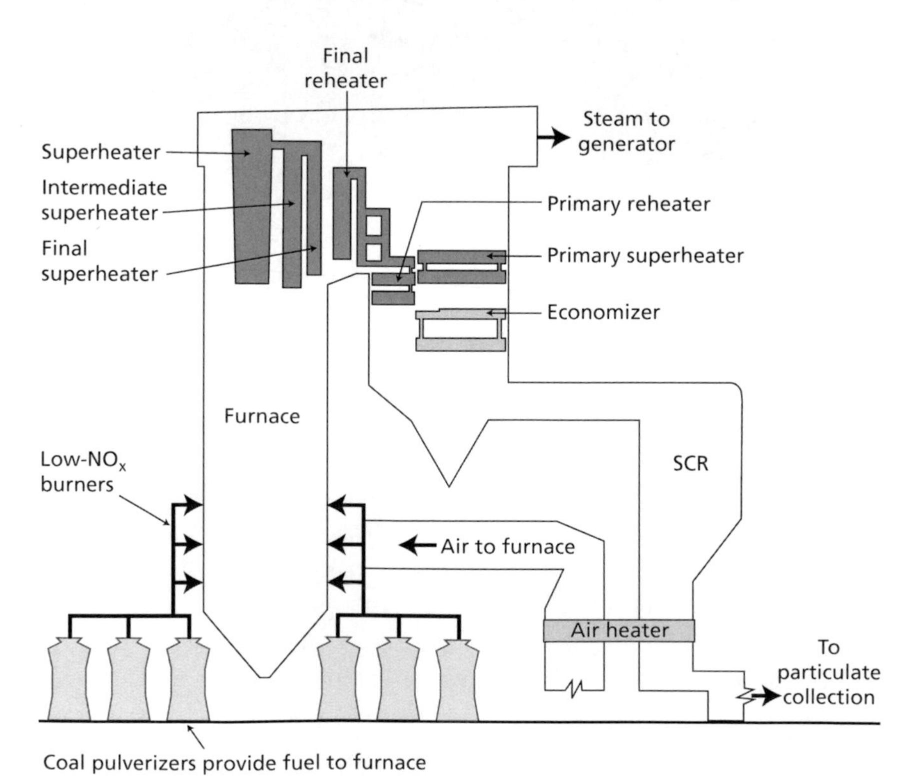

SOURCE: Kitto and Stultz, 2005.
RAND *MG1147-3.3*

lenges is a primary driver of boiler design, size, and arrangement, with lower-quality coal requiring larger boilers to reduce slagging (Kitto and Stultz, 2005). Ash content of coal can vary by region and within local areas, with most U.S. coal for power generation having an ash content between 6 and 20 percent (Kitto and Stultz, 2005). Chinese coals generally have more ash, with an average ash content of more than 23 percent (Sun, 2010). Boilers can be designed for multiple coal types, a process that results in increased costs and design trade-offs. Designing for multiple fuels is especially challenging in a supercritical plant because the circulation system needs to accommodate a wider variation in heat-transfer rates (Kitto and Stultz, 2005).

Pollution-Control Systems

To reduce the amount of SO_2 from coal combustion and therefore meet pollution-control regulations, plant owners sometimes switch to a lower-sulfur coal but often install "scrubber" or flue-gas desulfurization (FGD) systems to chemically remove the sulfur from the flue gas before it exits the smokestack. In a wet FGD system, the primary type of scrubber installed in coal-fired units, a mixture of water and limestone or other material is sprayed inside a tower, and it removes the sulfur from the flue gas passing through the tower. FGD subcomponents can weigh between 50 and 150 tons each (Kitto and Stultz, 2005). The tower is typically constructed with steel alloys and lined with rubber, tile, or coatings to protect the tower from the corrosive operating environment (Kitto and Stultz, 2005). A system of pipes, pumps, and grinders prepares and distributes the limestone mixture, and solid wastes are removed from the process and either taken to a landfill or sold as gypsum for drywall or other materials. Dry scrubbing systems spray chemicals mixed with far less water into a tower to remove sulfur from flue gases. These are less efficient at SO_2 removal but require less water and other materials, require less maintenance, and have lower operating costs (Kitto and Stultz, 2005). Direct sorbent injection (DSI) is a newer method of SO_2 control in which chemicals are injected directly into the furnace after the SCR. This method is attractive because it eliminates the large costs, materials, and water typically required to construct and operate FGD systems. Interviewees stated that this method generally has unfavorable economics due to the high cost and amounts of the injected chemicals required, but DSI is becoming a more attractive option for units that are required to reduce water use.

Other pollution-control systems at a coal-fired plant are not unique to coal-fired power, but they require specialized professional services to design and integrate into the power plant system. These include fabric filters and precipitators to reduce particulates, powdered activated carbon to reduce mercury, other methods to reduce hazardous air pollutants, water treatment and filtration, and solid-waste treatment, handling, and storage systems (Kitto and Stultz, 2005).

Smokestacks

After pollutants are reduced from the coal-combustion process, the remaining flue gases are exhausted to the atmosphere through a smokestack or chimney. The chimney is several hundred feet high, and chimneys are currently constructed from reinforced concrete and lined with fiberglass-reinforced polymers, borosilicate blocks, or a nickel-metal alloy (Anderson and Maroti, 2006). These chimney liners use specialty materials, are customized for this specific application, and are designed to resist the temperatures and corrosive environments in coal-fired power plant chimneys. Fiberglass-reinforced polymer liners are manufactured in sections using a filament winding process and are bonded together inside the chimney at the construction site (Bogner, 2009).

Implications of Variations in Coal Characteristics

Across and within countries and regions, the characteristics of coal can vary widely. These include its energy density, mineralogical and chemical composition, ash content, grindability, abrasiveness, erosiveness, moisture content, and other characteristics (Kitto and Stultz, 2005). Each coal EGU is designed to accommodate a range of coal types, yet capabilities in designing and operating units with different coals are critical in the industrial base. Equipment manufacturers maintain databases and design histories of projects across coal types and are gaining new experience with coals in foreign markets. Interviewees confirmed that U.S. coal-specific knowledge would be maintained even during a prolonged period of low activity in new coal power generating unit construction, primarily due to the maintenance of existing coal units.

Yet, the differences between U.S. and foreign coals present some industrial research and development (R&D) challenges. Because the long-term implications with operating some coal unit components with foreign coals are unknown, equipment manufacturers have difficulty predicting and designing to reduce material failure with foreign coals. Perhaps more importantly, experience gained using local coals in China and other countries deploying advanced coal generation technologies, such as ultrasupercritical, IGCC, and oxy-combustion, will provide some technology experience, but additional expertise on the design implications of using various U.S. coals with these technologies will be needed. These remain critical areas in R&D for advanced coal power generation.

CHAPTER FOUR

Market Structure

The coal-based electricity-generation equipment industry can be divided into three market sectors, distinguished by the type and scope of activities:

- new-unit construction
- O&M
- pollution control.

The new-unit construction sector includes the design and construction of new coal EGUs, either on an existing site or on a new site, and the design, development, and production of the major subsystems and components of coal EGUs (pulverizer, boiler, steam-turbine generator system, pollution control, and cooling system). After a coal EGU is constructed, on-site staff and contractors perform routine activities associated with unit operations, as well as a broad range of maintenance activities that can include simple repair or routine maintenance of unit subsystems to replacement or refurbishment of major components or subsystems. As discussed in Chapter Two, many existing units do not have advanced SO_2 controls. The pollution-control sector includes installation of subsystems to clean or capture waste streams resulting from the combustion of coal on existing units in response to changes in environmental regulations. Maintaining a complete coal-based electricity-generation industrial base requires sufficient activity in the new-unit construction market sectors. However, activity in the O&M and pollution-control sectors exercises some of the same skills and materials as the new-unit construction sector.

The New-Construction Sector for Coal-Fired Electricity-Generating Units

Constructing a new coal EGU requires the unit to be designed, components to be manufactured, and the components to be assembled at the site location. Labor and equipment are sourced differently in the new-unit construction sector. Nearly all the engineering design work for U.S. units is sourced domestically. According to interviewees, most of the coal-unique equipment (large boilers) and associated components (heavy-wall seamless tubing) for coal EGUs are manufactured outside the United States, while pollution-control equipment is manufactured domestically. Labor for construction and integration of a new unit is also nearly always sourced domestically.

Activity in the new-unit construction sector can be tracked with one of two metrics: the number of units installed or the amount of capacity installed, in megawatts. A unit generally consists of a boiler and steam-turbine generator system and can range in capacity from less than 150 MW up to about 1,300 MW, as discussed in Chapter Two. Each unit installed requires design, specification, manufacturing, and construction services, with larger units requiring additional time, materials, and labor (Kitto and Stultz, 2005).

Figure 4.1 shows the number of new coal EGUs brought online, by year, in the United States. A clear downward trend in new construction is apparent beginning in the mid-1980s. The number of new units installed reached ten per year in about 1987 and, except in three years, has remained below ten per year through 2010. Figure 4.1 also shows the capacity brought online, in megawatts and by year. Overlaying these data makes apparent that the industry shifted from more and smaller-capacity units in the 1950s and 1960s to fewer and larger-capacity units in the 1970s and 1980s. Over the course of the decline in installations in the past two decades, both small- and large-capacity units have been installed. Approximately 20 percent of the units installed between 2000 and 2010 have had a capacity of less than 50 MW (Ventyx, 2011).

Figure 4.1
Number of Currently Operating Coal Units and Capacity Installed in the United States, by Year

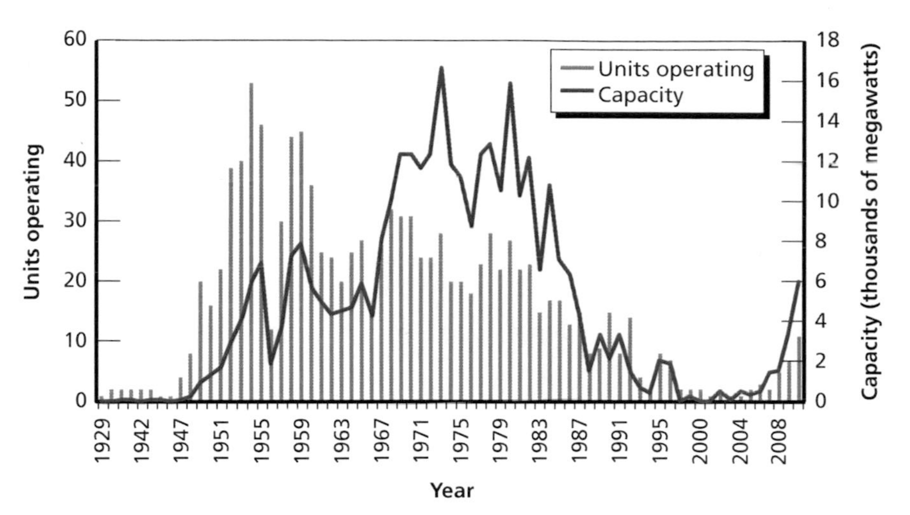

SOURCE: Ventyx, 2011.
RAND *MG1147-4.1*

Demand for New Coal Electricity-Generating Units

Eleven new coal EGUs totaling approximately 6 GW in installed capacity were brought online in 2010.[1] Nearly half of this capacity utilized supercritical designs. These new units are listed in Table 4.1, along with the EPC and boiler manufacturer. In contrast to the relatively weak demand in the United States, international demand, especially in China but also in India and non–Organisation for Economic Co-operation and Development (OECD) Asia, has been increasing. Since 1990, coal EGU installation activity in China has consistently been more than 10 GW annually, with a peak of more than 80 GW in 2006 (EIA, undated), as shown in Figure 4.2.[2] This activity in China

---

[1] Eleven are shown in the table; a twelfth, the 704-MW Brame Energy Center in Louisiana also came online in 2010; however, it is using petroleum coke as its primary fuel (Ventyx, 2011).

[2] Although Figure 4.2 shows all conventional thermal power capacity additions in China, coal dominates this sector, with natural gas comprising 5 percent of installed capacity. See EIA, 2011b.

Table 4.1
Coal-Fired Units Brought Online in 2010

Unit Name	Nameplate MW	State	EPC	Boiler Manufacturer
Oak Grove Steam Electric Station	878.6	Texas	Fluor Corporation	Babcock and Wilcox Company
Comanche Generating Station	856.8	Colo.	Shaw Group	Alstom
Iatan Generation Station	850	Mo.	Burns and McDonnell; Kiewit Corporation	Alstom
J. K. Spruce Power Plant	820	Texas	Zachry Group, Black and Veatch, Industrial Company, Utility Engineering	Alstom
Oak Creek Power Plant	615	Wis.	Bechtel Corporation	Hitachi Power Systems
Trimble County Generating Station	834	Ky.	Bechtel Corporation	Doosan Babcock
Plum Point Energy Station	720	Ark.	Black and Veatch, Zachry Holdings, Kiewit Corporation	IHI Corporation
John Twitty Energy Center (formerly, Southwest Power Station)	300	Mo.	Boldt Company	Foster Wheeler
Wygen III	100	Wyo.	Babcock and Wilcox Company	Babcock and Wilcox Company
Willmar Municipal Utilities	Four (two 2-MW units)	Minn.	Not applicable	Babcock and Wilcox Company

SOURCES: Ventyx, 2011 (megawatts and states); Industcards, undated; Spring, 2009; Peltier, 2008, 2011; Bechtel Corporation, 2006; Doosan Power Systems, 2010, 2011; Black and Veatch, 2010; NAES Corporation, 2011; "Southwest Power Station Unit 2 for City Utilities of Springfield," undated; Babcock and Wilcox, 2008, 2011.

is up to an order of magnitude higher than the 0–6 GW of annual installations in the United States. Of the approximately 800 GW total

Figure 4.2
Annual Fossil Power Plant Additions in China

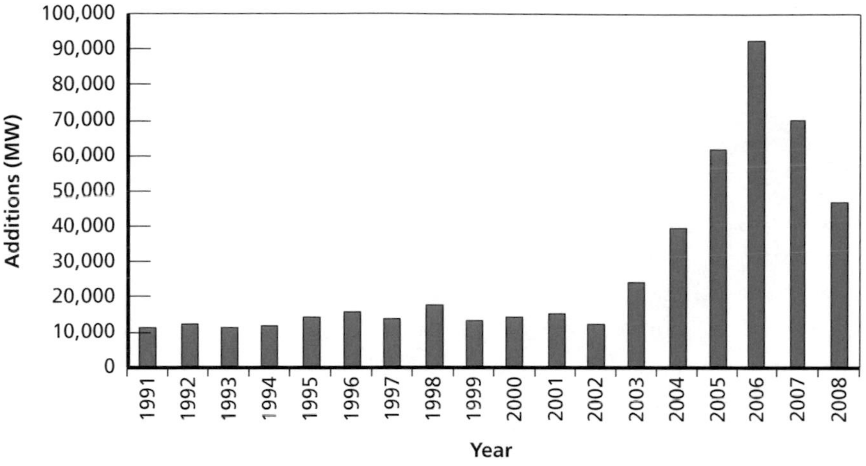

SOURCE: EIA, undated.
RAND *MG1147-4.2*

of new coal plants forecast globally for 2015–2035, nearly 600 GW is forecast in China, 50 GW is forecast in India, and 90 GW is forecast in non-OECD Asia (EIA, 2010a). Equipment costs and availability (lead times for manufacturing) in the United States are highly sensitive to this global market demand for these components and raw materials. Yet, labor supporting U.S. demand for new coal-unit construction is sensitive largely to U.S. demand rather than to a global market.

New-Unit Specification, Design, and Project Development

A utility plans for new capacity additions by conducting a series of feasibility studies, cost and risk assessments, and efficiency and load growth forecasts. Once the utility has established that new capacity is needed, it evaluates fuel, operating, and capital costs of generating options. If coal-fired power is the most suited to the utility's needs, it generally announces its intent to construct a coal EGU and begins the application process for regulatory permits to construct and operate a unit at the preferred site. A further, detailed feasibility study might also be conducted. If the state has a regulated electricity market, an application is made to the public utility commission to justify the spreading of the project costs across future retail electricity rates. Technology choice can also be affected through public utility commission discussions. Utilities typically allocate sufficient time for the permitting process to be completed and approved. In the months that follow a coal EGU's announcement, owners sometimes cancel the project due to the results of the feasibility study or because of difficulties in permitting. For projects that progress toward construction, often the permit's terms guide the design, technology, and specifications of the unit. According to requirements dictated by the permits, the owner finalizes the specifications of the desired coal EGU. These specifications include such elements as steam flow, temperature and pressure, boiler efficiency and emission requirements, type of fuel used, and scope and schedule of work (Kitto and Stultz, 2005). The utilities advertise or invite proposals from different AEC/EPC firms, evaluate bids, and select a vendor to design and construct the coal-fired power unit. The selected vendor develops an integrated master schedule and begins engineering design, long-lead procurement activities with equipment manufacturers, and preliminary site work.

The design of large, capital-intensive power plants involves specialized engineering knowledge and experience. Although this is a unique skill set, many U.S.-based AEC/EPC vendors exist with these capabilities. For the 11 units that came online in 2010, at least seven separate AEC/EPC firms were used. Every year, the trade publication *Engineering News-Record* surveys and ranks the top engineering firms by specialty and revenue. Of the 25 firms listed for fossil power plant design, 18 had more than $50 million in revenue in this market sector in 2010 ("The Top 500 Design Firms," 2011). Additionally, ten of the top 25 fossil-fuel power-plant design firms are also ranked in the top 20 for power construction contracting firms ("The Top 400 Contracting Firms," 2011). Table 4.2 presents the top ten design firms for fossil power generation from these rankings, showing the broad market capabilities that currently exist.

Our interviews revealed that U.S. offices of AEC/EPC firms and equipment manufacturers also typically conduct engineering and design for U.S. units. However, these U.S. offices often also conduct

Table 4.2
Top U.S. Fossil-Fuel Power Plant Engineering Design Firms

Firm	2010 Revenues in Fossil-Fuel Power Generation (millions of dollars)
Black and Veatch	358
Kiewit Corporation	192
Fluor Corporation	180
Burns and McDonnell	171
AECOM Technology Corporation	145
CH2M HILL	143
Bechtel	138
Zachry Holdings	134
URS Corporation	134
KBR	130

SOURCE: "The Top 500 Design Firms," 2011.

engineering and design of units to be installed internationally, which helps maintain these critical capabilities during periods of limited activity in the domestic market. The U.S. engineering and procurement teams work closely with the equipment manufacturing teams to communicate the exact specifications and schedules required for each project, yet AEC/EPC firms are rarely located close to equipment manufacturers.

Component Manufacturing. A coal EGU is comprised of a series of fabricated components that are manufactured offsite, shipped to a construction site in major subassemblies, and then assembled and installed on-site. Each coal EGU unit is customized to meet design requirements for power output, coal type, local environmental regulations and conditions, and other factors. Major components for coal EGUs are manufactured specifically for a project in response to an order for a new unit. Manufacturing these components utilizes specialized equipment. Consequently, lead times between order and delivery for major components are long. During periods of high demand, competition for manufacturing capabilities leads to higher prices or stretched-out delivery times.

The most critical system in the coal EGU component supply chain is the one that produces the pressure components of the boiler system. These include headers, drums, tubes and economizers, and other components designed to withstand high steam temperatures and pressures. The components are manufactured by rolling or pressing steel plates through hot or hollow forgings, piercing solid rounds of steel to produce seamless tubes, and bending and welding components into subassemblies for shipment. Many thicker components also require heat treatment to improve strength, which can require capital-intensive equipment for large components. Tubes, headers, and other components can be shipped to the construction site individually or can be joined into a complex subassembly for shipment and modular construction. By increasing the amount of final assembly labor conducted in the factory, considerable cost and schedule savings can be achieved compared with assembly on the construction site (Kitto and Stultz, 2005).

Welding is a critical function in the coal-fired electricity industrial base. Several welding methods are utilized throughout EGU component manufacturing. Continued advances in high-strength, high-temperature alloys for components have required the development of new welding techniques (Kitto and Stultz, 2005). The various welding types have differing skill-level requirements, quality, productivity, and cost. Electric arc welding, which can occur in several ways, is the most common welding process used in the manufacturing of coal-fired power unit components, although other methods are also used. Some welding in the manufacturing process is amenable to automation with machines or robots to improve quality and costs, but many welds are still performed manually (Kitto and Stultz, 2005). Table 4.3 outlines the attributes of several welding methods used in coal generating unit equipment manufacturing, construction, and repairs. Many of our interviewees expressed concern about the availability of welders trained to work on the high–steam pressure sections of the boiler, and these skills are a critical part of the coal-fired electricity industrial base.

Table 4.3
Welding Processes Used in Coal-Fired Power Plant Component Manufacturing, Construction, and Repairs

Weld Method	High Capital Costs	High Operator Skill Required	Readily Automated	Relevant Power Plant Component
Gas tungsten arc welding		x	x	Superheater, economizer, tubes
Gas metal arc welding			x	Tubes, panel wall repairs
Shielded metal arc welding		x		Tube-to-header walls, repairs
Submerged arc welding				Limited to header and water wall fabrication
Plasma arc welding	x			Sootblower tubes
High-frequency resistance welding	x			Tube manufacturing
Laser-beam welding	x			Overlay of tubes and panels

SOURCE: Kitto and Stultz, 2005.

In our interviews, a consensus emerged that nearly all pressure-component manufacturing currently occurs outside of the United States, mostly in Asia, with some facilities in eastern Europe. Several reasons were given for the current location of pressure-part manufacturing:

- desire to locate manufacturing supply closer to the centers of demand for coal EGUs
- lower labor costs and productivity gains achieved in international locations
- lack of available manufacturing facilities in the United States with steel heat-treatment capabilities suitable for very large EGU components

- need for U.S. AEC/EPCs to globally source major equipment in order to remain competitive in the U.S. market.

Some of these factors were also mentioned during our interviews in the context of why many other EGU components besides pressure parts were also manufactured in Asia. Because delivery times to the construction site are established early in the project development process, sufficient time exists for custom fabrication and shipment from an international supplier.

Table 4.4 lists some of the key suppliers of major components of coal-fired power units, compiled from lists of currently completed projects and from discussions with interviewees. In addition, Appendix A provides further detail on firms that provide AEC services, boilers, generators, pollution-control systems, and pulverizers. There are several things to note:

Table 4.4
Key Firms in the U.S. Coal-Fired Industrial Base

Contribution to the Base	Participating Firms
AEC/EPC	AECOM Technology Corporation[a], Bechtel[a], Black and Veatch[a], Burns and McDonnell[a], CH2M HILL[a], Fluor Corporation[a], Foster Wheeler, KBR, Kiewit[a], Shaw Group[a], URS Corporation[a], Zachry Holdings[a]
Pulverizers	Alstom, Babcock and Wilcox[a], Babcock Power[a], Columbia Steel Casting[a], Foster Wheeler, IHI Group
Burners	Advanced Combustion and Process Controls[a], Alstom, Babcock and Wilcox[a], Babcock Power, Coen Company[a], Damper Design[a], Foster Wheeler, Fuel Tech[a], RJM Corporation[a]
Boilers	Alstom, Babcock and Wilcox[a], Babcock Power[a], Doosan, Foster Wheeler, Hitachi Power Systems, IHI Group, Metso, Toshiba Plant Systems and Services
Turbines and generators	ABB, Alstom, Babcock and Wilcox[a], Foster Wheeler, GE Power[a], Hitachi Power Systems, Siemens, Toshiba Plant Systems and Services
Pollution control	ABB, Babcock and Wilcox[a], Babcock Power[a], BASF, Cormetech[a], DuPont[a], Foster Wheeler, GE Power[a], Haldor Topsoe, Hamon Group, Pullman Power[a], Siemens

[a] Firm headquartered in the United States.

- Many firms are capable of manufacturing key components (or, in the case of the EPCs, supplying a service), so purchasers enjoy some competition among manufacturers of these components and can choose among several possible vendors. Competition holds down costs and provides an incentive for technological innovation. The fact that there are many firms manufacturing the same component indicates a broad industry base.
- As demonstrated by the fact that a firm can appear in more than one category, some firms currently manufacture more than one component. Alstom, Foster Wheeler, and Babcock and Wilcox manufacture four or five of the key components. Although these components are all in the coal-based (or, more generally, conventional) electricity-generation industry, these firms do not need to rely on demand for a single component to maintain financial viability. For instance, manufacturing (and installing) pollution-control equipment for either new construction or retrofit on existing units is an important source of revenue when demand is low for other coal-unique components, such as large boilers.
- Many of these firms and, in particular, the market leaders for some components are not headquartered in the United States. Even for firms not headquartered in the United States, a common trait among large equipment manufacturers is performance of much of the global engineering design work in the United States. Hence, coal-based electricity-generation capability is supported by both U.S.- and foreign-based firms. This is an important aspect of the global market. International demand for coal-based electricity-generation equipment helps U.S.-based firms remain viable, while U.S. demand for that equipment can draw from the larger global marketplace.

Table 4.4 is not a complete list of firms capable of manufacturing one or more of these components or of providing AEC/EPC services. It includes the predominant firms producing large, utility-size boilers and generators. However, considerably more firms can provide AEC/EPC services or manufacture specific pollution-control subsystems. These firms are supported by a secondary tier of vendors that provide specific components and services for each of these major subsystems. We did not dig deeply into the secondary-tier vendors in this study, but they do constitute a critical element of coal-based electricity-generation equipment manufacturing capability. According to industry interviews, AEC/EPCs prefer to receive bids from at least three secondary-tier suppliers for major subcontracts; it is generally not difficult to reach this goal.

Some of the firms listed in Table 4.4 manufacture the same or similar components for the nuclear power industry: GE, Toshiba, Hitachi, and Alstom (D'Olier et al., 2005). As noted earlier, both the financial viability and experience base for the coal-based electricity industry should be evaluated within the context of the broader electric power generating equipment industry. Currently available data do not allow us to estimate how much of the coal-unique equipment manufacturing or manufacturing and construction workforce experience is directly transferable (overlap in knowledge and skills required), but the consensus among the interviewees for this study was that a large degree of fungibility exists between the natural gas, nuclear, and renewable generation workforces. This provides both a source of additional labor during shortages and a source of competition for similar resources during surges.

The coal-based electricity-generation industry is mature; coal-fired power plants similar to those currently being constructed have been generating electricity in the United States on a large scale since the middle of the 20th century. The fact that the industry is mature means that many technologies and processes are well understood and even somewhat standardized. However, the structure of the industry and, therefore, its capabilities have not been static. Figure 4.3 illustrates this by tracking the evolution of the U.S. boiler industry, an important subset of the overall coal-based electricity-generation industry. Boilers have coal-unique characteristics and, as discussed, are largely custom made to the specifications of a specific power unit. Prior to 1960 (the beginning of rapid growth in the industry), more than ten different U.S.

Figure 4.3
Evolution of Firms with Coal-Fired Boilers Installed in U.S. Power Plants in 2010

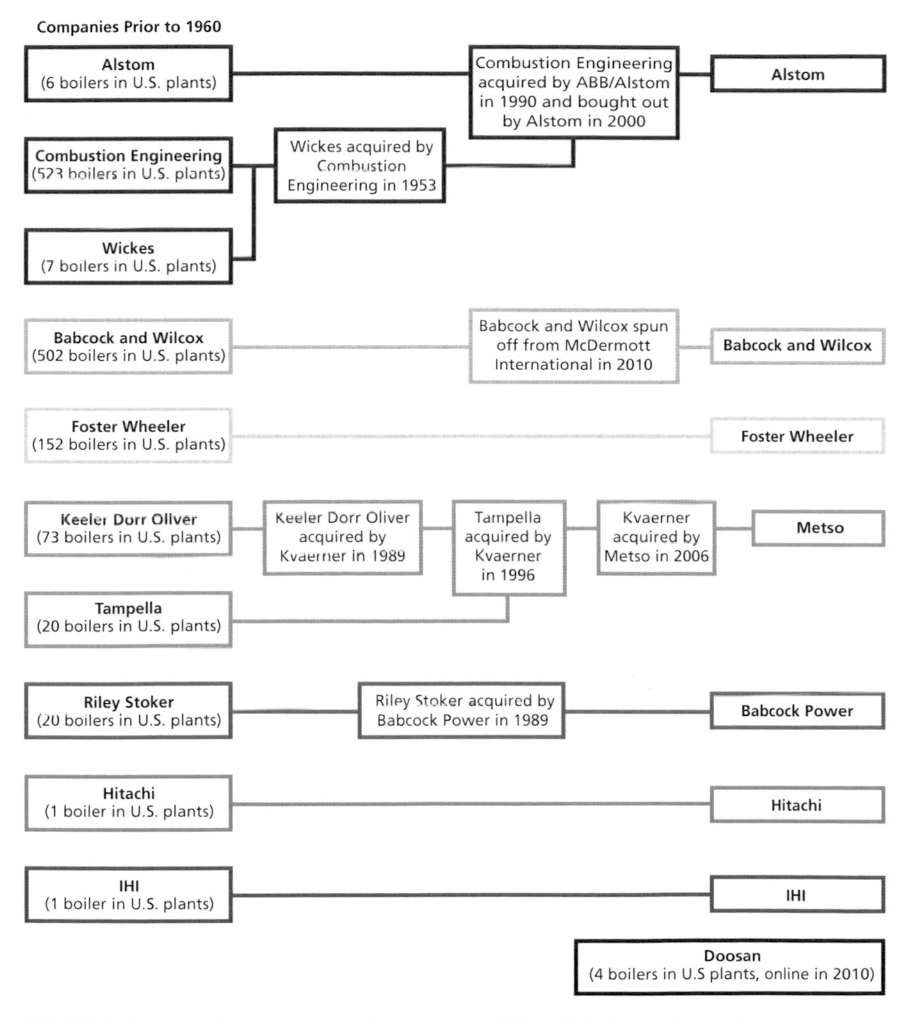

SOURCES: Information on companies prior to 1960 and their numbers of boilers was generated from NETL, 2007. Information on mergers and acquisitions from Prevost, 2011; ABB, 2000; "Wickes Companies," 1992; McDermott International, 2010; and Metso, undated.
NOTE: This does not list every company manufacturing boilers for coal-fired power plants. Smaller firms are not included.
RAND *MG1147-4.3*

companies manufactured boilers.[3] However, three firms dominated the market; Combustion Engineering, Babcock and Wilcox, and Foster Wheeler accounted for more than 90 percent of coal boilers installed in the United States. These same firms still dominate the U.S. market for operating boilers in coal-fired power plants. However, through mergers, acquisitions, and market changes, the number of large U.S. boiler firms has fallen. ABB bought Combustion Engineering and formed a joint venture with Alstom, and Alstom bought out ABB's share of the boiler business a year later. Earlier firms Keeler/Dorr-Oliver and Tampella eventually became part of Metso, and Riley Stoker was acquired by Babcock Power.

As discussed earlier, Babcock and Wilcox, Alstom, and Foster Wheeler account for the vast majority of the existing boilers installed in the coal-fired power plant fleet. However, new international entrants, such as Doosan, Hitachi, and IHI, have won recent contracts and compete for U.S. market share. In the U.S. market, the evolution has been toward more international firms. It is important to note that a firm's decision to locate its headquarters in the United States or in other countries depends on many factors and that foreign firms often rely on its U.S.-based workforce for engineering or other services.

Construction of Coal Electricity-Generating Units

Construction of an EGU begins after the procurement and construction schedule is developed. Preliminary construction work proceeds concurrently with ordering and fabrication of major, long-lead power-unit components. The EGU typically begins commercial operation about four years after construction commences; some units take longer. Many resources used during new-unit construction are similar to those used for other traditional large construction projects—skilled, unskilled, and professional labor, heavy construction equipment, cranes, and construction materials. This implies access to a broad supply chain for labor, equipment, and material, as well as competition with other construction projects for these inputs.

[3] Figure 4.3 is not a complete list of companies manufacturing boilers for coal-fired power plants; smaller firms are not listed in the figure.

A project management team and specialized skilled labor with experience in constructing large, capital-intensive coal-fired power units are critical to the coal-fired electricity industrial base. Construction engineering teams also support the construction managers and provide configurations for crane placements, logistics plans, structural analysis, and other services (Kitto and Stultz, 2005). Our interviewees discussed the fact that owners demand construction management teams with previous experience in large coal-fired power generating unit construction and that prior international experience is acceptable to satisfy this requirement.

The initial construction tasks involve preparing the site and building the unit's foundation. Then, the facilities, buildings, and structural work of the power unit begins, followed by equipment delivery, assembly, and installation. Finally, systems are connected, tested, and evaluated prior to the start of commercial operations.[4] An illustrative schedule of activities and workforce over the course of a coal EGU construction project is depicted in Figure 4.4.

Total unit costs are comprised of equipment, material, labor, AEC/EPC services, support facilities, and contingencies (Gerdes et al., 2010). NETL's 2010 estimate for a new PC supercritical coal plant is $1,647 per kilowatt. The Energy Information Administration's (EIA's) updated 2010 costs are $2,844 per kilowatt for a dual-unit plant and $3,167 per kilowatt for a single-unit plant (EIA, 2010c). NETL generated the cost estimates using vendor quotes and scaled estimates from previous projects. Its estimate includes equipment, supporting facilities, labor, engineering and construction services, and contingencies (Gerdes et al., 2010). Figure 4.5 shows the approximate share of total plant costs each system represents for a completed coal-fired plant. Figure 4.6 provides further detail on the distribution of total plant costs by category and system. As evident from the figures, equipment costs represent nearly half of total plant costs, and the boiler and NO_x-control system is the most expensive single system. Although increases in construction labor expenses increase total plant

---

[4] Construction-progress photographs of a recent coal power plant can be viewed at Prairie State Energy Campus, 2011.

Figure 4.4
Illustrative Coal-Fired Power Plant Construction Schedule, by Task and Number of Workers

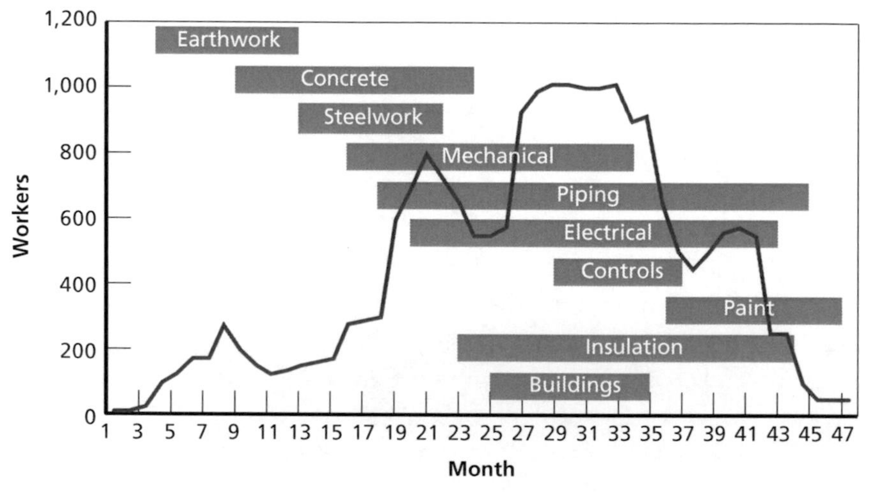

SOURCE: U.S. Department of Agriculture (USDA), 2007.
RAND *MG1147-4.4*

costs, the relative proportion of labor and equipment implies that total plant costs are more sensitive to increases in equipment costs than to changes in local labor costs.

Operation and Maintenance Sector

The set of activities comprising O&M is performed on the entire coal electricity fleet. Some skilled workers, trained technicians, and operators are required as full-time staff at coal power plants. These staff members manage fuel delivery and feed, boiler, feedwater, turbine or generator, cooling-tower pollution control, and electrical-control operations. According to our interviewees, older units generally require more staff than newer units do, and larger units require more staff than smaller units do. Operations staff members often undertake routine maintenance as needed in the power plant. Scheduled and unscheduled maintenance undertaken by off-site staff or contractors can vary

Figure 4.5
Distribution of Total Plant Costs, by Category

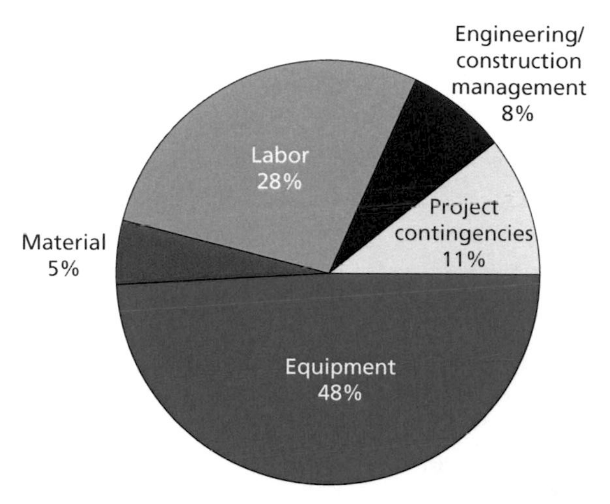

SOURCE: Gerdes et al., 2010.
NOTE: The data depict components of a supercritical 550-MW PC-fired plant.
RAND *MG1147-4.5*

from routine maintenance to the repair or replacement of large components. The market for these services is therefore large. Most of the current coal-fired generating units came online between 1965 and 1987 (see Figure 2.3 in Chapter Two), resulting in a fleet with a majority of units between 25 and 45 years old. Figure 4.7 shows the average annual fixed O&M expenditures by year of initial operation (i.e., age), which sums to an annual potential market of approximately $9.7 billion.

Depending on the plant, routine maintenance is largely performed in-house by the utility's operational staff or by domestic local service contractors. Larger projects can be performed by the OEMs. These projects take from days to weeks to complete, but very large upgrade and refurbishment projects can take months.

Examples of maintenance activities include the following:

- tube and pressure vessel leak repair
- fan and pump repair or replacement
- steam-turbine generator blade replacement
- boiler header and panel replacement

- FGD leak repair
- chimney crack and corrosion repair.

According to several interviewees, there is no "replacement and upgrade" market sector in which major subsystems (e.g., boilers, pulverizers, steam-turbine generators, boiler feedwater system) are replaced or upgraded completely. Rather, components of these major subsystems are repaired, replaced, or upgraded as parts wear out. This work is undertaken in the form of extended maintenance projects. Labor and equipment for these larger maintenance projects can draw on labor and equipment from both the U.S. domestic and international marketplaces.

Figure 4.6
Distribution of Total Plant Costs, by Category and System

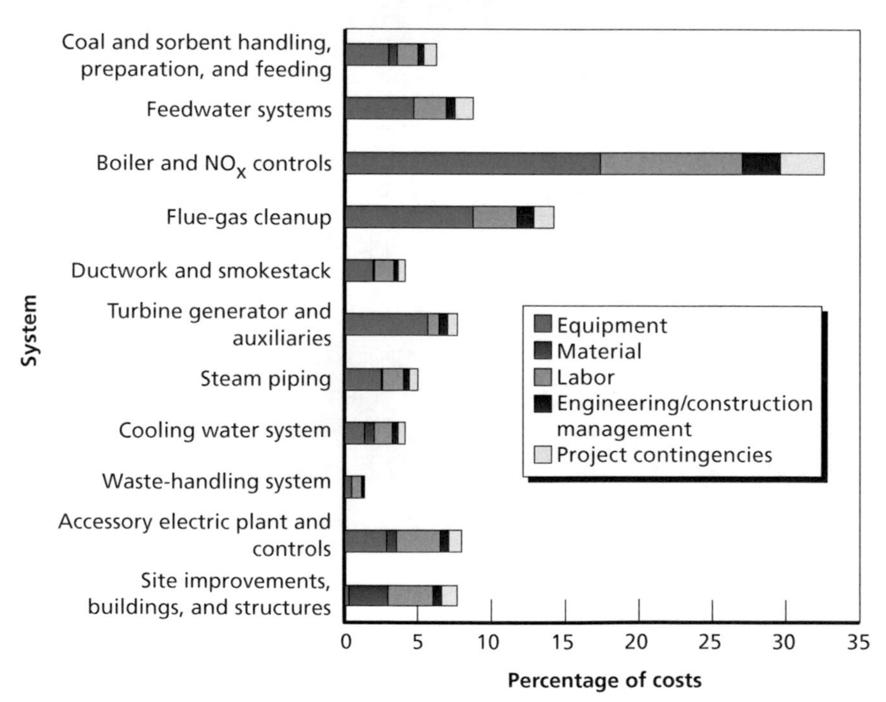

SOURCE: Gerdes et al., 2010.
NOTE: The data depict components of a supercritical 550-MW PC-fired plant.
RAND *MG1147-4.6*

Figure 4.7
Average Annual Fixed Operation and Maintenance Expenditures

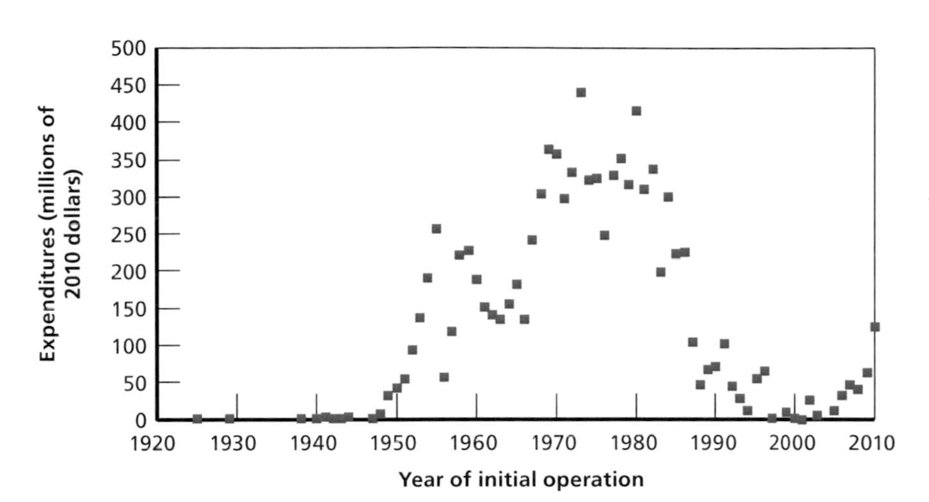

SOURCE: Ventyx, 2011.
RAND *MG1147-4.7*

Given the age of the fleet, this market appears to be substantial. It draws on and exercises most of the skilled labor required to build an entirely new coal EGU. Interviews suggest that engineering (i.e., design work) for these large projects is generally done domestically, while a significant portion of equipment manufacturing is done internationally and then shipped to the United States. Installation generally uses the labor pools of the OEMs or local service contractors. The breadth of activities in the O&M sector provides some of the experience base required to sustain a coal-based electricity-generation capability in industry. The major experience omitted in this sector is the type that leads to integrated design, specialized project management, scheduling, and procurement skills associated with constructing a new coal EGU.

Pollution-Control Sector

The need to add subsystems for pollution control is driven by regulations—most notably, under the Clean Air Act (Pub. L. 88-206,

1963), in the form of emission standards for waste streams resulting from the use of coal to generate electricity. Prior to 1990, the Clean Air Act exempted older plants from the additional emissions standards from New Source Performance Standards and New Source Review regulations unless they underwent a major modification (Kitto and Stultz, 2005; Pub. L. 88-206, 1963). The 1990 Clean Air Act Amendments (Pub. L. 101-549, 1990) established several regulatory trading programs to reduce SO_2 and NO_x, and it established maximum-achievable control technology (MACT) regulations to reduce air toxins. Hence, each generating unit in the coal-fired fleet was designed and constructed under a specific regulatory environment, and, as a result, the coal-fired fleet contains a mix of units with varying levels of pollution controls installed. Increasing emission-control standards change the attributes of the required pollution-control equipment. As such, regulations imposing new emission requirements might require adding subsystems to existing units, and these additions define the market for pollution-control equipment manufacturing and installation to a large extent. The current or proposed regulations (e.g., Cross-State Air Pollution Rule reducing SO_2 and NO_x, which contribute to particulate and ozone pollution [EPA, 2011f]; mercury and hazardous air pollutants [EPA, 2011d]; coal-combustion residuals [EPA, 2010]; and water intake [EPA, 2011c]), as well as potential GHG regulations (EPA, 2011e), create an uncertain environment regarding their eventual implementation that directly affects investment in both pollution-control retrofits and new construction of coal-fired power units. Adding pollution-control subsystems to an existing unit requires a fairly large capital expenditure, which increases the cost of producing electricity.

Current regulations for coal EGU include NO_x and SO_2 emission standards, as well as standards for particulates. Under current law, new units must include the best available control technology for these pollutants. Major retrofits to existing units would include the installation of pollution-control technology.

The specific conventional pollution-control technologies, or subsystems, include the following (Kitto and Stultz, 2005):

- SCR to reduce NO_x emissions
- powder-activated carbon (PAC) to reduce mercury emissions
- electrostatic precipitator (either wet or dry) to reduce fine particulates
- fabric filers to reduce particulate matter
- FGD (wet or dry; often referred to as a scrubber) to reduce SO_2 emissions
- DSI to reduce SO_2, sulfur trioxide (SO_3), and other acid gases
- low-NO_x burner.

With the exception of the burner, these are all postcombustion subsystems and so can theoretically be retrofitted to an existing unit. Burners, more-efficient boilers, or advanced control systems can also be retrofitted to existing units.

The potential for future standards, either under the Clean Air Act or through other legislation, controlling GHGs from EGUs causes perhaps the greatest uncertainty (EPA, 2011e). Coal-fired EGUs account for approximately 80 percent of electricity-sector carbon dioxide (CO_2) emissions and 25 percent of total GHG emissions in the United States (EIA, 2010b). CO_2 emissions from coal-fired EGUs can be reduced by improving combustion efficiency (which can provide incremental improvements), utilizing combined-heat-and-power applications (which provide electricity and useful heat that is otherwise lost), or by separating and storing CO_2 through precombustion (e.g., via IGCC), postcombustion (e.g., via chemical reaction with a solvent), or oxycombustion (via combustion in relatively pure oxygen) CCS (Interagency Task Force on Carbon Capture and Storage, 2010). All CCS technologies are still in development to scale up for utility applications, with only limited demonstration to date (Interagency Task Force on Carbon Capture and Storage, 2010).[5] Because of the additional equipment and electricity required for operation, adding CCS to existing units is expected to significantly increase the cost of electricity gener-

---

[5] For example, a 20-MW portion of an existing coal unit in West Virginia is capturing and storing CO_2, as a technology validation, but the utility decided to suspend further project development. For a list of global CCS projects, see Carbon Capture and Sequestration Technologies Program, undated.

ated from coal (Interagency Task Force on Carbon Capture and Storage, 2010), and research to reduce these costs through research, demonstration, and deployment is ongoing (NETL, undated).

There have been several recent analyses of the effect of emerging or anticipated regulations on the coal-fired fleet. Most of these focus on the decision to either retrofit or retire existing units as a result of new regulations. Results are driven by assumptions regarding the specifics and timing of the regulations and the cost of retrofits in comparison to alternative energy sources (renewables, natural gas, and efficiency improvements).[6] These assumptions lead to both significant uncertainty and a wide range of results. Differences in source data also contribute to differences in results. However, such estimates do provide estimates of the size of the pollution-control market sector. Listed here are the summaries of some of the several studies examining these issues:

- Celebi et al. (2010) estimates that up to 55 GW of coal-fired power plant capacity would retire as a result of tightened NO_x and SO_2 standards. It estimates costs at $70 billion to $130 billion for the 187 GW of coal-fired capacity required to comply. Although merchant power plants would account for most of the retirements, regulated plants would account for most of the compliance costs.
- Wood Mackenzie (2010) estimates that up to 60 GW of installed capacity would be retired in the next ten years (double the rate of the previous decade). The new mercury standards would require capital investments of about $100 per kilowatt; a unit with a nameplate capacity of 600 MW would thus cost $60 million to retrofit.
- Bernstein Research (2010) finds that capital costs to add scrubbers to existing units vary by unit size, with a 50-MW unit costing $1,137 per kilowatt, a 400-MW unit costing $470 per kilowatt, and a 1,000-MW unit costing $358 per kilowatt.

---

[6] Some analyses also distinguish between merchant- and privately owned plants, which rely more on revenues from electricity sales to recoup investment, and regulated utilities, for which cost recovery can be built into the rate structure. See, for instance, Celebi et al., 2010.

- Credit Suisse (2010) estimates 50 GW in coal-fired fleet retirements, with another 100 GW requiring extensive pollution-control retrofitting that could cost from $70 billion to $100 billion.
- ICF/Edison Electric Institute finds 25 GW of coal retirements by 2020 in its reference case without any new air, water, or ash regulations. When alternative policy cases are analyzed, it finds that between 17 and 76 GW of incremental retirements will occur.
- The Bipartisan Policy Center finds that air, water, and waste regulations will increase coal-unit retirements in 2030 by 15 and 21 GW relative to base-case retirements, with more retirements occurring assuming low natural gas prices. It argues that the industry has demonstrated the ability to simultaneously install pollution-control equipment across a considerable portion of the fleet, citing the 60 GW of capacity retrofitted with scrubbers between 2008 and 2010.
- Bradley et al. (2011) find that the industry can meet the proposed regulations without threatening system reliability, predicated on proper planning and implementation.

Figure 4.8 shows the level of activity in pollution-control installation for SO_2 reduction.

As these data show and as interviewees have conveyed, the pollution-control sector is currently a fairly active market. As noted earlier, many of the key firms that specialize in manufacturing coal-unique components (boilers, heavy-wall tubing) also manufacture some pollution-control equipment. In contrast with boiler manufacturing, for instance, pollution-control manufacturing still takes place in the United States. Several firms noted that their pollution-control divisions helped keep them viable during downturns in demand for the more specialized components. Large pollution-control retrofit projects typically span 15 to 60 months and require specialized and integrated design, fabrication, and construction services to execute (Kitto and Stultz, 2005). Much of the pollution-control technology can be considered unique to coal-fired power plants. According to interviewees, few other industrial processes require the same type and level of pollution-control subsystems.

Figure 4.8
Sulfur Dioxide Scrubber Installation Activity, by Year

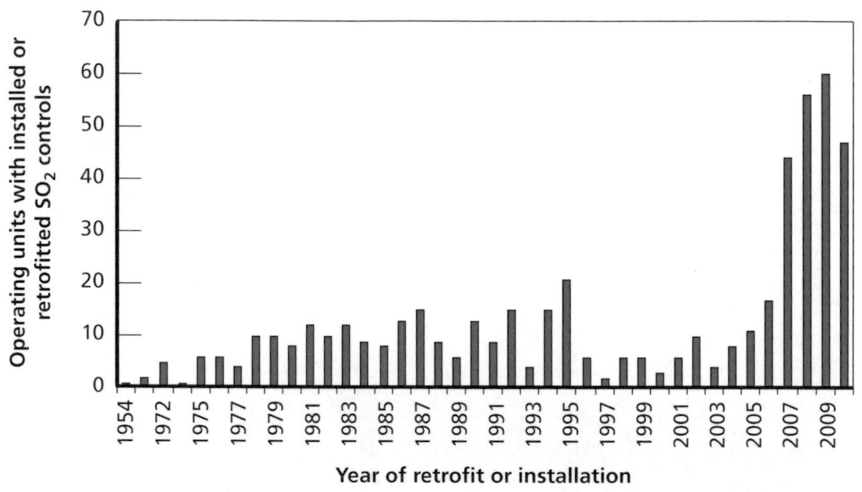

SOURCE: Ventyx, 2011.
RAND *MG1147-4.8*

Proposed new regulations are likely to increase activity in this market sector. The current market size is estimated to be around $5 billion per year, according to interviewees. However, about 140 GW (40 percent of the capacity) and 700 units (60 percent of the units) do not have FGD equipment. Similarly, about 50 percent of the capacity and 65 percent of the units have no SCR or SNCR equipment for NO_x reduction (Ventyx, 2011). Responding to these gaps, as well as proposed regulations on mercury and other toxic air emissions, ash, water, and GHGs, could require considerable investment. Much of the required pollution-control equipment is manufactured in the United States. Some firms have been relying on revenue from the pollution-control market to maintain technical and financial viability.

New or changing regulations are likely to continue to drive demand within this market sector, resulting in a potentially large market for firms manufacturing and installing pollution-control equipment. Complying with these regulations generally means adding one or more specific pollution-control subsystems to the back end of an existing plant to capture and treat specific emissions. Thus, some of the activities unique to front-end processes and equipment (boilers, heavy-wall seamless tubing, generators) are not necessarily exercised. In addition, the decision of whether to comply with new or changing emission standards by adding or enhancing specific pollution-control subsystems will be made in a larger context that includes the cost of compliance, future demand for electricity, and the cost of alternative sources for that electricity. In some circumstances, a utility might choose to temporarily close or permanently retire an existing coal EGU rather than incur the cost of compliance with new or changing regulations.

Another point that was often made in our interviews and that is also made in other industry reports is that a surge in demand for pollution-control equipment could overwhelm existing industry capabilities. This situation could occur if all existing coal-fired units are ordered to comply with new or changing regulations within a small, fixed window of time. Interviewees expressed concern that the near-term demand and short compliance timelines will substantially increase construction costs and schedules for pollution-control retrofits and, in tandem, for new-unit construction. After the installation of pollution-control retrofits on existing units is complete, this market will be limited to pollution controls for new–coal unit construction. Hence, interviewees were concerned that incentives for pollution-control firms to add capacity during the compliance period would be limited. EPA's analysis of previous FGD installations and use of project acceleration assumptions results in estimates of capabilities for 70 GW of cumulative wet FGD installations by 2015 or 2016 to comply with the regulation of hazardous air pollutants, depending on the rate of increase in capabilities during the compliance timeline (EPA, 2011b). This is an area that should continue to be examined for potential bottlenecks and strategies to alleviate cost and schedule pressures.

Summary of Findings on Market Structure

The market for coal-based electricity-generation equipment is global. The United States no longer dominates in terms of demand for equipment or coal EGU construction. The U.S. industry has contracted from historical highs, and foreign firms now compete in the U.S. market. The U.S. market for new coal EGUs is on the order of 2–6 GW per year, substantially lower than during the peak years of the 1970s and 1980s. The manufacturing base for pressure-part components is now largely outside of the United States, though several firms maintain core engineering design capabilities within the United States. The influence of the international market, both on supply and demand, is increasingly important to firms based in the United States. These changes represent industry's response to market signals and changes in the business environment.

The primary components in a coal EGU that are unique to coal-based electricity generation are the pressure-part components. These include boilers, large forgings (such as boiler headers), and heavy-wall seamless piping. These components are largely no longer manufactured in the United States. However, the U.S. market has access to the handful of firms that are capable of producing these components. These firms include a few U.S.-based firms that have manufacturing activities overseas.

One of the major challenges the industry currently faces is the uncertainty associated with new regulations and emission standards for SO_2, NO_x, particulates, ash, mercury, HAPs, water use, and CO_2. The regulatory structure and emission standards affecting electricity-generation technology and fuel-source choices is very important in the United States. In general, environmental regulations tend to make the price of electricity from coal higher than that of other fuel sources. The technology for CCS has been demonstrated only on a small scale and is therefore immature. According to some interviewees, the uncertainty surrounding near-future environmental regulations adversely affects industry's investment in capital and the workforce. Although installation of SO_2 scrubbers on existing units will continue to exercise the pollution-control sector, uncertainty regarding other regulations (most notably, potential CO_2 regulations) has caused utilities to reconsider large investments in new coal plants. The recent drop in overall electricity demand (related to the 2008 recession) and competition from low-cost natural gas power generation significantly affect the outlook for near-term investment in new U.S. coal-fired generating capacity (EPA, 2011f).

However, the U.S. industrial base is comprised of interconnected market segments: new-unit construction and equipment manufacturing, O&M on the existing fleet, and pollution-control additions or upgrades to the existing fleet. The O&M and pollution-control markets appear to be fairly robust, even while new construction–sector activity is sporadic and lower than it was in the 1970s and 1980s. These sectors exercise some, but not all, areas of the U.S. industrial base. Manufacturing of coal-unique equipment occurs largely outside the United States, but engineering design and manufacturing of pollution-control equipment remains largely within the United States. Thus, the United States is not at risk of losing the capability to design, construct, operate, and maintain a coal-based electricity-generation capability.

These findings are consistent with the findings of a 2005 U.S. Department of Energy assessment of the infrastructure required to design and construct a nuclear power plant. In particular, that assessment concluded that the "necessary manufacturing, fabrication, labor, and construction equipment infrastructure is available today [2005] or can be easily developed to support construction and commissioning of up to eight U.S. nuclear units during the period from 2010 to 2017" (D'Olier et al., 2005). Further, the assessment noted that the nuclear power industry was now global, resulting in some competition for resources. Additional long-lead planning was required for some critical components for which there are few (or only one) certified vendors. Even after several decades without new–nuclear unit construction in the United States, the assessment concluded, the capability to respond to an increase in demand still existed.

CHAPTER FIVE

Workforce

A sufficient workforce of individuals with the skills and experience required to design, manufacture, construct, operate, and maintain coal EGUs is a critical component of industry capability. Some aspects of workforce issues were addressed in Chapter Four in the context of some segments of the market. But there are also more-general workforce issues that warrant a separate discussion.

There has been a series of reports examining the energy-related workforce, motivated in part by the perception that the current workforce is aging, a significant portion of the workforce will retire in the near future, and fewer new hires are available to replace them (see, for instance, U.S. Department of Labor, 2007; U.S. Department of Energy, 2006; Bipartisan Policy Center, 2009; Center for Energy Workforce Development, 2009; EPRI, 2007). Most of the studies focus on the workforce associated with constructing and operating power plants and maintaining power lines; the workforce associated with manufacturing key subsystems and components has received less attention.

An examination of national employment data in these sectors can help provide context for the power industry as a whole. Figure 5.1 shows changes in the numbers of four categories of workers relevant to the coal-based electricity-generation industry. Compared with a 1990 baseline, the workforce has been steadily declining in the boiler-manufacturing industry and the fossil-fuel electric power generation industry, in contrast with increases in the workforce associated with power-line construction. The short-term variation in power-line construction and, to a lesser extent, turbine and turbine generator manu-

Figure 5.1
Trends in the Utility-Related Workforce

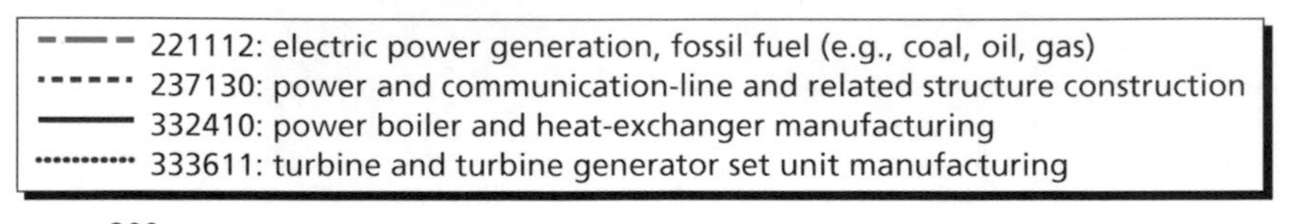

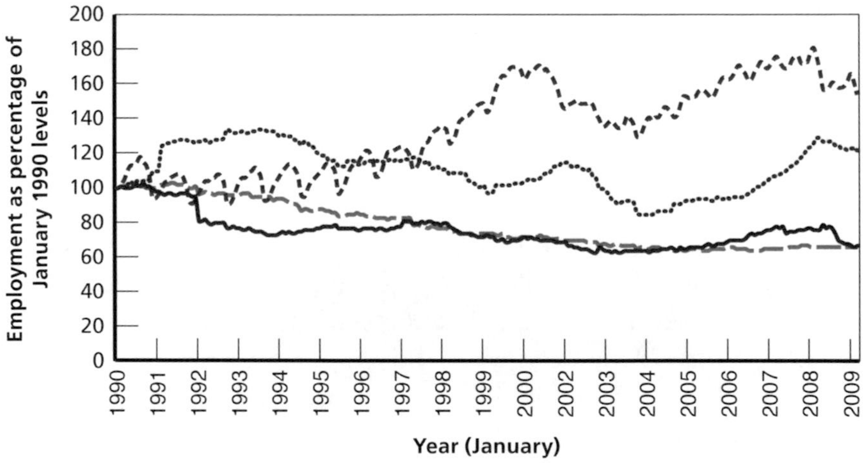

SOURCE: Bureau of Labor Statistics (BLS), undated.
RAND *MG1147-5.1*

facturing reflects seasonal fluctuations in the workforce. As is typical of the national data in this industry, the data we use are not detailed enough to directly reflect workforce trends associated specifically with coal EGUs or to provide comparisons to other sources of electricity generation; yet, overall downward trends are noted in the fossil-fuel electric power generation sector and power boiler– and heat exchanger–manufacturing sector.

Table 5.1 shows employment levels in certain key occupations and skilled trades associated with the construction, O&M, and retrofitting of coal EGUs. Once again, data do not exist specifically for coal EGUs. However, existing data do illustrate employment in the construction of power plants and employment by utilities and thus represent an upper limit on those involved with the construction of coal EGUs. Table 5.1 also shows that the number of workers who are either involved in utility-system construction or employed directly by utilities

Table 5.1
Workforce Snapshot in Relevant Occupations for 2009 and Percentage Change from 2003 to 2007

Occupation	National Total	Heavy and Civil Engineering Construction	Utility System Construction	Utilities
Boilermaker	22,400	3,010	2,600	190
Carpenter	743,760	27,220	6,660	260
Cement mason or concrete finisher	165,700	17,290	3,300	70
Operating engineer or other construction-equipment operator	368,200	109,550	45,010	3,530
Electrician	579,150	7,010	5,340	8,680
Insulation worker: floor, ceiling, and wall	26,500	60	0	0
Insulation worker: mechanical	29,620	1,500	1,130	0
Plumber, pipefitter, or steamfitter	400,970	18,170	16,670	8,600
Reinforcing iron or rebar worker	24,200	2,520	360	0
Millwright	41,640	1,840	1,340	100
Laborer or freight, stock, or material mover, hand	2,135,790	10,090	5,070	3,500
Construction manager	204,760	22,910	8,760	770
Engineering manager	178,110	1,430	360	4,630
Civil engineer	259,320	8,810	1,470	1,890

Table 5.1—Continued

Occupation	National Total	Heavy and Civil Engineering Construction	Utility System Construction	Utilities
Civil engineering technician	82,690	1,000	200	700
Construction laborer	856,440	184,470	79,390	3,020
Structural iron or steel worker	65,130	6,490	2,350	0
Construction or related worker, all other	47,630	2,250	1,220	190

SOURCE: BLS, undated.

NOTE: All industry categories are subsets of national data and are mutually exclusive from one another. Employees are all part-time and full-time workers who are paid a wage or salary. The survey does not cover the self-employed, owners or partners in unincorporated firms, household workers, or unpaid family workers.

is relatively small across all occupations and skilled-labor categories in comparison to the national totals.

Two observations can be drawn from the small number of workers employed in fields associated with the construction of EGUs, a number that is small relative to the national total. First, to the extent that skills are transferable between industries, there is a large pool of potential workers for power units. That is, if a pipefitter for a coal EGU has the same skills and qualifications as a pipefitter in another industry, there are many potential workers from which to draw. Similarly, the skills and qualifications of pipefitters, electricians, or boilermakers might be fungible to some extent across electricity-generation technologies, such as coal and nuclear power. Although specific skill sets within an occupation are not necessarily common to all workers in that category (i.e., not all boilermakers are familiar with the procedures for assembling and installing the large boilers used in coal EGUs), the construction and utility market sectors are drawing from a larger workforce pool.

Second, to the extent that the individuals employed in construction of EGUs can be substituted by those from other industries, the labor market for these skills will be affected only in small ways by trends in the coal EGU industry. Trends in employment and wages are more likely to be affected by larger, macroeconomic trends; costs of materials; and general demand for these skills. Conversely, the workforce associated with the construction, O&M, and retrofit of coal EGUs can likely maintain some level of coal-specific competency, even if they occasionally work in other economic sectors.

However, these data do not offer any insights into whether these skills are transferable. To consider the specialty of any laborer, there are two relevant criteria:

- How similar is the skill required for coal EGUs to the skill required for other projects (different generating units, industrial construction, or other fields)?
- How large is the market for the other projects that are related to these skills?

If, for example, pipefitters engaged in the construction of coal EGUs can be used at other large industrial construction projects, and the reverse is also true, then this skill does not depend greatly on trends and changes in the coal-fired electricity industrial base. However, some skill sets require experience on coal EGUs. Here, ongoing experience with coal EGUs is highly valuable, according to interviewees. These skills are a unique competency. For these very specific skills, the national market for the occupation does not provide complete context and insight.

The New-Construction Workforce for Coal Electricity-Generating Units

Labor is a critical input in the construction of coal EGUs. Table B.1 in Appendix B provides the number of individuals employed nationally in key occupations associated with utility-system construction for several recent years. The occupation list is similar to that in Table 5.1. We also show data for the two construction-related industries: utility-system construction and other heavy and civil engineering construction.

The data provide a mixed message. For example, the plumber, pipefitter, and steamfitter occupations show a decrease of about 33,000 workers between 2003 and 2009. The number of workers in this occupation in the other heavy and civil engineering construction industry also shows a decrease. However, the utility-system construction industry—the industry that appears to be closest to coal EGU construction—shows substantial growth in employment.

There are limited data on the workforce size associated with specific coal EGU construction. The 1,600 MW Prairie State Energy Campus has employed approximately 3,500 workers during peak construction activity (Prairie State Energy Campus, 2010). This workforce included the following craft positions: boilermakers, bricklayers, carpenters and millwrights, cement masons, electricians, elevator constructors, glaziers, insulators, ironworkers, laborers, operating engineers, painters, pipefitters, roofers, sheet-metal workers, and teamsters. This list of occupations is similar to that shown in Table 5.1. Basin Electric's 385-MW Dry Fork Station employed approximately 1,300 workers during its peak construction period (Basin Electric Power Cooperative, undated).

Estimates, as described earlier, range from between 1,300 to 4,000 individuals involved in construction of coal EGUs at any one time. The Ochs Center for Metropolitan Studies collected employment information on six coal-fired power plants constructed between 2005 and 2009 with capacity greater than 500 MW. The study cited estimates that power plant construction would increase employment by 1,000 to 2,400 people. However, by looking at county-level data, the center estimated that, in most cases, actual, permanent employment was smaller than initial projections made prior to construction (Ochs Center for Metropolitan Studies, 2011).

We note that the county-level estimates do not necessarily represent the number of individuals working at the power plant. County-level employment might have increased by less than the amount projected if individuals employed on other projects simply shifted to the power plant. The county-level employment data might overestimate the number of individuals involved with construction if other construction projects were peaking at the same time. However, these estimates help demonstrate the range of personnel required for construction. For a Bipartisan Policy Center report, Bechtel estimated the number of individuals with specific skill required in the construction of a coal EGU (see Figure 5.2) (Bipartisan Policy Center, 2009). These projections are sensitive to some assumptions and parameters and should not be used as precise estimates of the workforce required for construction of a coal EGU. However, these estimates are consistent with estimates cited earlier and with data presented earlier.

Although credible estimates exist for the workforce required to build a coal EGU, national-level data are not granular enough to present information about the number of people with specific skill sets (i.e., the number in any particular occupation) required for construction of coal EGUs.

Figure 5.2
Bechtel's Illustrative Estimates of Skilled Labor for Coal Power Plant Construction

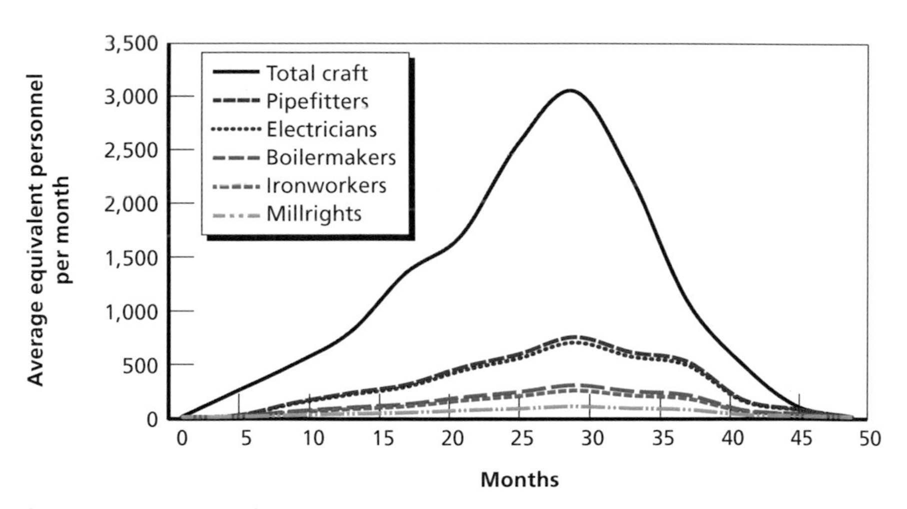

SOURCE: Bipartisan Policy Center, 2009.
NOTE: The figure presents illustrative estimates of skilled labor per gigawatt of generation for a supercritical PC plant, based on 600-MW blocks. Data shown are critical-craft average equivalent personnel per month, subject to a ±25-percent variance in estimate.
RAND *MG1147-5.2*

Equipment-Manufacturing Workforce

Table B.2 in Appendix B provides data on the workforce supporting equipment manufacturing. These data from BLS show occupations relevant to coal-based electricity-generation equipment manufacturing between 2003 and 2009.[1] The occupation-level data can also be shown by industry. We show two industries that include parts of coal-based electricity-generation equipment manufacturing: boiler, tank and shipping-container manufacturing, and engine, turbine, and power-transmission manufacturing. The coal-based electricity equipment of interest in this study is a subset of these industry categories—specifically, boilers and turbine generators. Unfortunately, there are no data that allow us a more detailed and precise view of the workforce in which we are interested. These broader occupation and industry numbers provide scale and allow researchers to examine trends at a higher level of aggregation.

Interviews across firms stated that much of the equipment manufacturing has moved abroad. This is particularly true for manufacturing boilers; almost no boilers are currently manufactured in the United States for coal EGUs. As the physical location of the manufacturing has moved abroad (primarily to China, Japan, Korea, Thailand, and eastern Europe), the U.S. workforce has moved to other available occupations. Interviews indicate that there were two primary reasons the industry has shifted outside the United States:

- Labor is cheaper outside of the United States. This has been a macro trend across many manufacturing sectors in the United States for years and holds true in the coal-fired electricity equipment industry.
- The market for coal EGUs is an increasingly global marketplace; there is little U.S. demand for equipment for coal-fired power units. Thus, barring large increases in logistics costs (or risks), market forces affecting domestic demand and manufacturing

[1] Note that the 2009 national total number for each occupation corresponds to the number shown in Table 5.1.

costs do not support major investments in domestic manufacturing capabilities to support new-unit construction.

As shown in Figure 5.1, the number of workers in the power boiler and heat-exchanger subsector has decreased steadily for the past 20 years, which is consistent with information stating that these jobs are moving abroad. However, available data cannot confirm that these jobs are being replaced abroad (rather than just disappearing completely) or that the labor that has left is associated directly with coal EGUs rather than boilers for other purposes. In fact, the occupation-level data in Table B.1 indicate that boilermakers have seen a slight increase in employment in recent years.

However, although some manufacturing has moved abroad, interviews indicate that significant, but not all, engineering capabilities involved in the design of this equipment remain in the United States. There were no publicly available data to confirm or reject this claim.

Available data do not provide specific information on the coal-fired power plant industry. Any information in tables includes data on other industries, so no definitive conclusions can be drawn. Despite these caveats, they do provide some observations:

- The size of the workforce was relatively steady between 2003 and 2009, suggesting that the trends stated in the interviews began prior to 2004.
- The United States continues to have a substantial construction engineering workforce.

Some occupations that are critical to the coal-fired electricity industrial base—boilermakers, pipefitters, and welders—show a relatively flat trend. Other occupations show a decrease between 2003 and 2009. In addition, the data for 2008 and 2009 likely reflect the significant decrease in all manufacturing-related activities experienced in the United States in the economic downturn of the late 2000s.

Operation and Maintenance Workforce

A full-time workforce is required to operate the power plant and provide minor maintenance, while external vendors can be contracted to conduct annual and major maintenance and refurbishments. For one recently constructed plant, the company estimated that about 75 full-time employees would be required to operate the plant (USDA, 2007). These positions include managers, foremen, electricians, operating engineers, mechanics, and instrumentation and control technicians. Subcontracted maintenance personnel include boilermakers, pipefitters, electricians, and other skilled labor used in power plant construction. Table B.3 in Appendix B shows occupations related to the O&M workforce. These occupations correspond to the list in Table 5.1. We also show the breakout for the utility industry because this industry contains the coal EGUs of interest here. The total and utility data shown in Table B.3 are the same as those shown in Table 5.1 for 2009; we have added data covering 2003–2008.

By necessity, the O&M workforce is domestic. Figure 5.1 suggests that the past 20 years have seen a decline in the workforce associated with fossil-fuel power generation. We do not know whether the same trends have been experienced by the coal-fired electricity-generation industry.

Table B.3 in Appendix B shows numbers of workers for utilities, which might approximate those working in an O&M role. However, the table does not provide specific data about the coal-fired electricity-generation industry. As in the case of manufacturing, these data suggest no major trends in workforce between 2003 and 2009. Most occupations were relatively flat over this period; a few increased (e.g., civil engineers), and some decreased (e.g., engineering managers, steel workers), but, in most cases, the change has been small.

Pollution-Control Retrofit Workforce

BLS data are not detailed enough to identify a workforce associated with pollution controls. It is likely that this workforce is included in some of our other data because a fair amount of the work is manufacturing pollution-control equipment, construction and installation of that equipment in the power generation sector, or O&M of that equipment by the in-house utility workforce.

Interviews suggested that manufacturing for pollution control has not followed the same trends as other equipment manufacturing; it is still domestic. No data were found to verify these statements. An estimate of employment impacts for a series of model scrubber installations forecasted that a scrubber for a large utility EGU would use 103–129 engineering full-time equivalent positions and 25–31 iron or steelworker full-time equivalent positions during the manufacturing stage (Price et al., 2011).

By necessity, the workforce for the installation of pollution-control equipment is domestic. As described in previous sections, considerable pollution-control retrofits have been undertaken in recent years, but no specific information was available to demonstrate trends within this workforce. The Price et al. estimate forecasts that the installation of a scrubber for a large utility EGU would use 510–638 boilermaker full-time equivalent positions during the installation stage (Price et al., 2011).

Interviewees expressed concern that the near-term demand and short compliance timelines involved with installing equipment to meet proposed U.S. Environmental Protection Agency (EPA) pollution-control regulations will substantially increase construction costs and schedules for pollution-control retrofits and, in tandem, for new-unit construction. After the installation of pollution-control retrofits to existing units is complete, this market will be limited to pollution controls for the new–coal unit construction. Hence, interviewees noted, incentives for pollution-control firms to add capacity during the compliance period would be limited. This is an area that should continue to be examined for potential bottlenecks and strategies to alleviate workforce, cost, and schedule pressures.

Workforce Demographics

Several interviewees said that a significant portion of the skilled-labor pool is nearing retirement. These statements were often combined with an observation that hiring and training for new skilled labor were well below the replacement rate. Unfortunately, we were unable to obtain data that would verify the workforce aging demographic for coal-EGU design, equipment manufacturing, construction, operation, and maintenance. However, a few organizations, including both the U.S. Departments of Labor (DOL) and Energy, produced reports that addressed challenges to the energy workforce (DOL, 2007; U.S. Department of Energy, 2006; Bipartisan Policy Center, 2009; CEWD, 2007, 2008, 2009). Although these reports distinguish between the workforce that builds and operates power plants and the workforce that focuses on transmission and distribution, they do not distinguish among different kinds of electricity-generation technologies.

According to the 2009 CEWD "Gaps in the Energy Workforce Pipeline," roughly half the workforce in several relevant job classifications will be eligible for retirement between 2009 and 2015, including 46 percent of pipefitters, pipelayers, and welders; 51 percent of technicians; and 51 percent of engineers (CEWD, 2009).[2] Perhaps more significantly, utilities responding to the survey reported having trouble filling openings in skilled labor positions: "between 30 [and] 50 percent of applicants (those [who] met the minimum requirements for a position) were not able to pass the pre-employment aptitude test." We heard this in several of our interviews as well.

Although these results are not specific to coal, they do indicate that the utility industry perceives a current shortage of skilled labor in the workforce and anticipates that the shortage will grow worse in the near future. Without surveying companies individually, both to determine the age of the workforce and to determine which professions are truly unique to the construction of coal EGUs, it is difficult to verify these concerns. The data in Figure 5.1, which shows decreasing labor associated with fossil-fuel power plants, are consistent with the concern that the workforce is declining and new workers are not being hired at a one-for-one rate of replacement, but they do not explain the causes of this decrease in the workforce. However, concerns about an aging

---

[2] The survey included 31 companies accounting for 44 percent of U.S. electricity and natural gas utilities. See CEWD, 2009.

workforce present a problem only if there is a surge in construction of coal-fired power units. That is, a declining workforce represents a concern only if the remaining workforce does not represent enough individuals to cover the number of units demanded.

In interviews, individuals expressed concern that, should there be an increase in construction of coal EGUs, there would likely be shortages of skilled labor that would increase labor costs and slow the construction of coal-fired generating units. Of particular concern to interviewees were the supply, productivity, and costs of skilled welders critical to construction of coal-fired generating units. However, interviewees were confident that the industry could respond to correct labor shortages if a period of sustained demand were to arise. There are several reasons that they believed this to be the case:

- With a global marketplace for coal-fired generating units, the technical knowledge of how to build a power plant will not be permanently lost with retirements.
- The markets for O&M and installation of pollution-control equipment, as well as construction of other electricity-generation technologies, such as natural gas–fired power plants, provide some level of relevant training for workers.
- If a sustained surge in demand occurred, firms might need to invest in training programs to increase the size and capabilities of the workforce, and the supply of workers could be increased to meet a sustained demand within a few years.

Hiring and Training

Interviewees made several points relating to workforce hiring and training that reflect concerns for the industry. Although we were unable to objectively validate these concerns, we believe that they should be considered in the context of maintaining the industry's capability with respect to designing, constructing, and operating coal-fired power units.

Several interviewees noted that it was increasingly difficult to attract students and potential new workers to the coal-fired electricity industry. The lack of stable career opportunities, as well as the need to travel to be close to major projects, is undesirable at current wages to many potential employees. Additionally, interviewees expressed that coal still connotes "dirty" in the public eye and that many of these jobs are physically difficult. As many as half of job candidates fail the entrance exam, something we heard from several interviewees, shrinking the pool of potential candidates.

We heard several examples of industry teaming with community colleges or trade schools to establish programs that lead to a range of skill sets or occupations needed in the coal-based electricity-generation industry. In general, skilled-labor occupations require two to four years of apprenticeship before an individual is considered capable of working alone.

Current Experience from Recent Projects

Many interviewees had concerns about the number and quality of people in welding and other skilled trades during the recent construction boom. Because of the high level of aggregation in the workforce statistics described earlier, these concerns were not able to be verified by the data. Hence, we searched for examples in recent project experience. As part of characterizing the cost growth experiences of recent coal power unit projects discussed in Chapter Two, we found that many projects exhibited cost growth but that this growth was caused primarily by price increases in materials and equipment because of global demand. We also examined the state public utility commission testimony of one utility executive discussing the sources of cost growth for a specific recent plant (Haviland, 2011). This project was the first investment of its kind at that size, an IGCC power plant. As such, unforeseen engineering challenges and changes, larger-than-expected material quantity requirements, and delays increased costs considerably. The testimony mentioned poor labor productivity as the project progressed, especially regarding specialty welding services. However, the testimony largely attributes poor labor productivity to the unanticipated numbers required and inadequate project and resource planning that resulted. Thus, workers often did not have the materials needed on-site for construction. This caused a ripple effect through the schedule and budget, pushing peak working needs into winter months, which further affected productivity and costs. The testimony also stated the unplanned increase in requirements for construction materials substantially increased the number of welders needed and that the use of specialty alloys in the project caused welding difficulties (Haviland, 2011). Therefore, the project about which he was testifying points to several potential causes for increased costs, but they are not clearly separable from learning issues associated with a project that is first of its kind at its size. The industry should examine the experiences at that IGCC plant to improve designs, schedules, and workforce planning for future advanced coal power plants.

Summary of Findings on Workforce

Potential shortfalls in the coal-based electricity-generation workforce were among the motivators for this study. Unfortunately, publicly available information provides only limited insight into the workforce. BLS data provide useful contextual information but do not go down to the level of detail required to understand workforce trends specific to coal-based utility generation. In any case, the BLS data send a somewhat mixed message. Although there have been some declines in employment in related industries and key occupations (skilled labor), in other industry categories and occupations, employment has been maintained or grown. The data suggest that the coal-based electricity-generation workforce is pulled from a much larger pool of workers with similar skills and experience. This would tend to mitigate adverse effects specific to coal-based electricity generation.

Other reports present aggregate data (i.e., for the energy industry) that suggest an aging demographic, with the implication that a significant portion of the workforce is or will be eligible for retirement in the near term. However, we could find no publicly available information that documents a retirement surge in the coal-fired electricity industry. We were also unable to find data that show a shortfall in specific skill sets, with associated impacts on recent or current construction projects.

In general, we found that the workforce responds to market forces in a way that is similar to the way it does in equipment manufacturing. Stakeholders share responsibility for the hiring and training of new workers and for the organization of existing workers, including utilities; trade unions; equipment manufacturers; architecture, engineering, and construction (AEC) and EPC firms; and community colleges and trade schools. In the face of a sustained increase in demand, according to their experience and interview discussions, these stakeholders are well-suited to collaborate to eventually generate the level of skilled labor needed. However, this adjustment phase is likely to be accompanied by increases in the cost of labor, albeit temporary, and a period of longer-than-normal construction times until efforts to increase the pool of skilled labor are implemented or demand subsides. One potential strategy for mitigating shortages of skilled labor is an increased emphasis on programs that increase awareness of career opportunities and provide training and apprenticeship opportunities, and several interviewees said that their organizations had active programs in these areas. There is potentially a role for federal support for these kinds of programs and partnership with the organizations that run them.

CHAPTER SIX

Observations and Remaining Questions

The U.S. industrial base that supports coal-fired electricity generation is part of a global market for materials, manufacturing capacity, and engineering, construction, and project management services. According to projections for new installations of coal-fired EGUs and discussions with representatives of industry sectors, this global demand is dominated by the growth of the power sector in China. Of the approximately 800 GW of coal EGUs projected to be installed globally between 2011 and 2035, more than 600 GW are expected to be in China (EIA, 2010a).

In comparison, the U.S. market recently has provided 2–6 GW per year of newly installed coal-fired generation capacity. Manufacturing capacity of critical components to support this growth has been supplied largely by overseas facilities, with the U.S. demand looking to Japan for high-pressure tubing and to China for large forgings, such as boiler headers and turbine rotors. However, the engineering, project management services, and skilled construction labor (especially pressure-part welding) was drawn from the U.S domestic suppliers.

Forecasts indicate that U.S. domestic demand for new coal-fired generation capacity will remain low, with some projections suggesting very few installations in the next decade (EIA, 2011a). If limited new coal power unit construction occurs in the near term, U.S. firms will need to look toward other sectors or other markets for business. Discussions with representatives across industry sectors suggest that both could occur.

Although manufacturing for the global power-generation equipment market is already largely located overseas, U.S. engineering and construction management firms are actively engaged in foreign markets. Foreign markets are especially significant in the engineering design sectors, in which U.S.-based design groups provide engineering services for foreign projects. This has allowed firms to maintain many of the skills and capacity for designing new coal-based power generation facilities in the United States.

In addition, new construction, maintenance, and pollution-control retrofit activity has drawn largely on the skilled, domestic construction workforce. Our interviews and reviews of recently completed construction projects reveal that, although skilled labor in critical tasks was less abundant than in prior construction booms, labor shortages were not the dominant contributor to schedule delays and cost overruns.

The power-generation equipment industry also experiences continued activity or growth in other sectors. The O&M sector is expected to continue to be stable for the next decade. The pollution-control industry is expected to grow significantly as utilities comply with expected regulations. Subsequently, demand will fall when most retrofits are completed. The O&M and pollution-control sectors exercise some aspects of engineering, skilled construction, and manufacturing capacities. However, these activities do not require the management of a project as significant and complex as a new generation facility. Aside from limited maintenance and replacement of aging boilers, the work does not require skilled labor involved in new generation projects (e.g., large demands for pressure-part welding and rigging for coal-fired power plant construction).

The surge in the construction of U.S. coal-fired power plants between 2004 and 2010 provides insights into the challenges that the industry faces and the industry's ability to maintain the full range of required capabilities. The study approach included examination of available historical data to describe how the industry responded during this time.

Modest skilled-labor and other shortages were revealed during this construction surge. International competition for raw materials raised the price of critical materials, such as steel and specialty alloys. Longer lead times were required to obtain large-scale forgings or the heavy-wall tubing required for coal boilers and associated systems. These factors increased the overall costs of coal-fired generation unit projects or, in more-extreme cases, caused the delay or cancellation of ongoing planning or construction activities.

However, as shown by the global cost indexes cited in Figure 2.6 in Chapter Two and evidenced by interviewee comments, increased cost for power plants is not strictly a U.S. phenomenon. The cost pressures of the recent past were driven by demand for construction materials and services in China and elsewhere, not solely by increased U.S. demand. The role of potential shortages of skilled labor in the United States is more difficult to ascertain in terms of cost pressures. However, the decline in costs in 2010 and high unemployment in construction trades show that these pressures are sensitive to overall market demand.

Implications for the U.S. Coal-Based Electricity-Generation Industrial Base

This monograph has documented substantial changes in the U.S. coal-fired electricity-generation industrial base since the 1970s and 1980s. Some key components of coal-fired power generating equipment are no longer produced in the United States. For cost and other reasons, production has migrated to eastern Asia and eastern Europe. However, other components continue to be manufactured in the United States, especially pollution-control equipment. U.S.-based design firms are active globally and remain highly competitive.

The United States maintains capabilities in installation and O&M. We saw no signs that the market has failed to provide components, construction and installation services, or the labor required to provide those goods and services. Utilities have been able to obtain the equipment and services they demand for the recent increased demand for new generation. When utilities have needed to secure skilled labor in periods of high demand, they have paid premiums to enter into preferential service contracts or worked with local training institutions to develop apprenticeship programs so that the requisite labor has been available.

At the end of the most recent boom, the industry experienced price spikes or stretched-out delivery times for specific components. However, these spikes appear to have been driven largely by high global demand for materials and equipment and were transient, returning to normal levels after demand fell or when industry could expand manufacturing capacity.

The preponderance of evidence examined for this monograph indicates that, despite the recent recession, the industry base to supply equipment for coal-based electricity generation remains capable of responding to potential demands in the near term. Activity and prices respond to trends in the domestic and global markets for labor, materials, and equipment. Interviewees from all sectors of the industry had similar perspectives on this score. Most sectors did express a desire for more-stable supply and prices, as opposed to cyclical booms and busts in schedules and prices. However, each submarket appeared to be making the trade-offs between adjusting to a fluctuating market and paying to maintain excess capacity during periods of slower demand.

Historical and current analogous conditions from other industries provide further support for the idea that industry maintains required capabilities. The issues that motivated this study on the coal-based electricity equipment industry are not new. An aging workforce and changes in technology, regulatory environment, and market demand have affected other U.S. industries, both within and outside the energy sector. The U.S. Department of Defense (DoD) has raised a similar set of concerns in the private-sector industries supporting some weapon-system categories, such as military aircraft (Drezner et al., 1992; Birkler et al., 2003) and ships (Schank, Arena, et al., 2007; Arena et al., 2005). The international shipbuilding industry has also been studied in the United Kingdom (Schank, Riposo, et al., 2005; Pung et al., 2008; Bassford et al., 2010).

Within the U.S. energy sector, the civilian nuclear power industry has experienced similar cycles, given that no new nuclear power plant has been constructed in the United States in many years. A 2005 report (D'Olier et al., 2005) that addressed the ability of the U.S. nuclear power industry to respond to a surge in demand for nuclear power plants reaches findings very similar to ours:

- The nuclear industry is now global; U.S. firms, along with their international competitors, respond to global market signals.
- The drop in demand for new nuclear construction forced firms to maintain their capabilities through O&M, including retrofits and upgrades of selected subsystems, as well as international demand for new construction.
- Although some critical nuclear components are still manufactured in the United States, many components are manufactured offshore. However, there are no shortfalls in availability for components, although, in some cases, longer lead times exist.
- Although there are no critical shortages in skilled labor for the nuclear industry, some additional planning is required to compensate for selected skilled labor in high demand (e.g., pipefitters, welders).

The report concludes that the infrastructure exists, or can be easily reconstituted, to support a surge in new construction of nuclear power plants. One strategy being employed by the nuclear power industry to mitigate potential skilled-labor shortfalls and high costs is modular construction, which involves assembly of large modules at the point at which components are manufactured. Modular construction techniques are also used in the coal-fired power equipment industry; for instance, boiler components are assembled and shipped to the plant site as modules. Following this model, it might be reasonable to conclude that the industry base to support coal-based electricity generation in the United States will be able to exist within a global market and reconstitute itself domestically as necessary to support future demand.

The analogies to other industries must be viewed within the context of conditions specific to those industries. For example, at least three factors present in the nuclear power industry and that differ from those in the coal-fired electricity-generation industry affect the ability to maintain capability:

- The U.S. Department of Energy has been actively involved in maintaining the nuclear infrastructure, including at least two analyses of the issue. The department is also responsible for maintaining the nuclear weapon infrastructure and so maintains a core government capability on nuclear issues.
- The U.S. Navy and two shipyards, General Dynamics Electric Boat and Northrop Grumman Shipbuilding—Newport News, maintain an infrastructure supporting nuclear power in ships (carriers) and submarines.
- The three U.S. private-sector firms with nuclear power plant design capability—GE, Westinghouse, and Babcock and Wilcox—have invested in advanced reactor designs.

The direct involvement of two large government agencies and the national security implications of nuclear power have combined to keep interest and visibility high in the nuclear power industry. These factors are not present in the coal-based electricity-generation industry.

Despite the anecdotal and analogous historical evidence that supports the assessment of a capable industry base, it is important to consider alternative perspectives that were present in these same data. Some industry participants questioned what effect a prolonged drought of a decade or more without construction of a single new coal-fired power unit would have on the availability of domestic skilled labor, especially in project management and engineering. Others raised questions of how the effects of these trends, or even the price spikes at the end of the recent boom, would affect the competitiveness of coal-fired power generation and thus have implications for the long-term costs of electricity generation if the capability to utilize domestic coal resources for electricity production became limited.

According to our findings of a dynamic and evolving industry, the United States will not lose the capability to construct a new coal-fired EGU. However, the industrial base that supplies coal-based electricity generation in the United States has been diminishing as domestic demand declines. As a result, any revival of demand for new–coal unit construction will likely involve higher costs and longer production schedules. These conditions could be expected to exist until either demand declines or the industry responds to increase the workforce and production capacity. Which happens first will depend on conditions in the global market and the size and duration of the demand for new coal-fired generation equipment domestically.

Remaining Questions

Given that this industry comprises a large proportion of the U.S. electricity sector and that advanced coal power generation with CCS is one of the strategic technologies for GHG emission reduction, further investigation of a few questions could be worthwhile. Understanding the capabilities of the coal-based electricity-generation industry for a future transition to advanced technologies requires improved information in three critical areas: the key firms active in the industry, the workforce size and skills needed to cover the full range of activities, and improving understanding of likely demands that will be placed on that sector in the future.

Although we have a fairly good picture of the key firms producing coal-unique equipment and services, we do not have sufficient details to properly assess their financial and technical viability. For instance, there are no publicly available data reflecting the proportion of revenue derived from coal-related activities as opposed to all the other energy- and non–energy-related activities in which these firms are active. In addition, our analysis focused on supercritical and subcritical coal power units and stayed largely at the top tier of the industry. To assess the capabilities of the coal power industrial base to respond to a transition to advanced technologies, a better understanding of the key firms across the supply chain providing critical components for these technologies would be necessary.

The type and amount of available data that described the numbers of skilled and professional workers to support the coal-based power generation industry make it difficult to draw conclusions from the observations described previously in this chapter. In general, specific workforce requirements for design and construction projects are considered proprietary, and statistics of the number of different types of workers in the labor force are collected and reported at too high a level of aggregation to support this type of industry-base study, as demonstrated by the BLS data in Chapter Five. Collecting these data would require conducting surveys of individual firms and thus would involve handling business-sensitive or proprietary data. Recognizing the need for such refined data, at least one industry trade organization is reported to be collecting such detailed information for its specific market sector. However, to characterize the industry, these data are needed across the full set of sectors and subsectors. Given the broad use and sensitivity of these data, there might be a role for the Department of Energy partnering with industry to collect and periodically maintain statistics on workforce requirements and capabilities.

One potential strategy for mitigating shortages of skilled labor is an increased emphasis on programs that increase awareness of career opportunities and provide training and apprenticeship opportunities. These programs frequently engage industry trade organizations, unions, community colleges, and trade schools. Several interviewees said that their organizations had active programs in this area. There is potentially a role for the Department of Energy in supporting these kinds of programs and partnering with the organizations that run them. The type of detailed workforce data described here would enable formation and management of such engagements.

Ultimately, the interest in the capabilities of the coal-based generation industry base is driven by questions of whether the industry will be able to respond to future demand. However, demand is highly uncertain and depends on forecasts of economic growth, coal power's competitiveness relative to other generation technologies, the success of technological development efforts to improve the efficiencies of and reduce the emissions associated with coal-fired and other power generation, and the path that future pollution and GHG regulations take.

Costs and project schedules are likely to increase when new coal power projects are initiated after a long period of low demand, during high-demand periods caused by short regulatory compliance timelines or market forces, and from resource and workforce competition from constructing other electricity-generation technologies. Hence, this monograph is a first step in understanding the implications of a large-scale power-sector reinvestment and transition to advanced technologies for the U.S. industrial base. Although this and other studies have examined workforce and equipment needs for specific generation technologies, a holistic analysis, including multiple technologies deployed across a range of technology, demand, and construction scenarios, is needed to anticipate gaps in skills and resources and to determine robust strategies to alleviate barriers to a large power-sector reinvestment.

APPENDIX A

Key Companies in the Coal-Based Electricity-Generation Industry

This appendix provides an expanded list of the firms providing critical components, subsystems, or services in the coal-based electricity-generation equipment industry from that provided in Chapter Four. It is not meant to be an exhaustive, comprehensive list of such firms. Rather, the list represents the information we could obtain from publicly available sources and our industry interviews. We focus on firms whose products are associated with the coal-unique aspects of the industry.

We have excluded both private and public utilities, though these organizations provide the market for coal-based electricity-generation equipment.

The purpose of these tables is simply to demonstrate that the industry supporting coal-based electricity-generation capability in the United States is both broad and deep in many of the key components. Most of these firms are large, as measured by the size of their workforce and the revenue they obtain from the power sector. The global market within which the firms operate means that their business base, workforce, and level of experience draw from the broader international marketplace, which allows them to maintain core competencies and capabilities better than if they relied on the U.S. market for coal-based electricity-generation equipment alone.

Table A.1
U.S. Architecture, Engineering, and Construction Firms in the Power Sector

Market and Company	2011 U.S. Rank in Fossil Fuel Power Design Revenues	2011 U.S. Rank in Power Construction Revenues	2011 U.S. Rank in Air Pollution Services Revenues	2010 International Power Design Revenue ($ millions)	2010 International Power Construction Revenue ($ millions)
Black and Veatch	1	12	10	177	135
Kiewit Corporation	2	4	Not applicable	25	511
Fluor Corporation	3	3	3	61	694
Burns and McDonnell	4	15	12	Not applicable	Not applicable
AECOM Technology Corporation	5	Not applicable	18	210	Not applicable
CH2M HILL	6	18	4	27	106
Bechtel	7	1	2	49	Not applicable
Zachry Holdings	8	10	Not applicable	Not applicable	Not applicable
URS Corporation	9	6	1	37	71
KBR	10	7	Not applicable	87	469

SOURCES: "The Top 400 Contractors," 2011; "The Top 500 Design Firms," 2011; "The Top 200 Environmental Firms," 2011; "The Top 200 International Design Firms," 2011; "The Top 225 International Contractors," 2011.

Table A.2
Selected Parent Companies of Original Equipment Manufacturers of Boilers

Firm	2010 Revenue ($ millions)	Total Number of Employees	Headquartered in United States
Hitachi	121,704	361,745	
Alstom	28,315	93,500	
Doosan	21,000	38,000	
IHI Group	15,500	26,035	
Metso	7,522	38,000	
Foster Wheeler	4,068	12,000	
Babcock and Wilcox	2,689	22,000	x
Babcock Power	Not available	Not available	x

SOURCES: Hitachi, undated; Alstom, undated; Doosan, 2010; IHI, undated; Metso, 2011; Foster Wheeler, undated, 2011; Babcock and Wilcox, 2011.

Table A.3
Selected Parent Companies of Original Equipment Manufacturers of Generators

Firm	2010 Revenue ($ millions)	Total Number of Employees	Headquartered in United States
GE	150,211	287,000	x
Hitachi	121,704	361,745	
Siemens	102,935	405,000	
Toshiba	83,600	202,638	
ABB	31,589	116,500	
Alstom	28,315	93,500	
Foster Wheeler	4,068	12,000	

SOURCES: GE, 2010; Hitachi, undated; Siemens, undated; Toshiba, undated, 2011; ABB, 2011; Alstom, undated; Foster Wheeler, undated, 2011.

Table A.4
Selected Parent Companies of Original Equipment Manufacturers for Pollution Control

Firm	2010 Revenue ($ millions)	Total Number of Employees	Headquartered in United States
GE	150,211	287,000	x
Hitachi	121,704	361,745	
Siemens	102,935	405,000	
DuPont	32,733	60,000	x
ABB	31,589	116,500	
BASF	25,011	110,289	
IHI Group	15,500	26,035	
Foster Wheeler	4,068	12,000	
Babcock and Wilcox	2,689	22,000	x
Haldor Topsoe	765	2,015	
ADA Environmental Solutions	22	54	x
Babcock Power	Not available	Not available	x
Marsulex Environmental Technologies	Not available	Not available	

SOURCES: GE, 2010; Hitachi, undated; Siemens, undated; DuPont, 2011; ABB, 2011; BASF, 2011; IHI, undated; Foster Wheeler, undated, 2011; Babcock and Wilcox, 2011; Haldor Topsoe, undated; ADA Environmental Solutions, 2011.

Table A.5
Selected Original Equipment Manufacturers for Pulverizers

Firm	2010 Revenue ($ millions)	Total Number of Employees	Headquartered in United States
Alstom	28,315	93,500	
IHI Group	15,500	26,035	
Foster Wheeler	4,068	12,000	
Babcock and Wilcox	2,700	29,000	x
Babcock Power	Not available	Not available	x
Columbia Steel Casting	Not available	Not available	x

SOURCES: Alstom, undated; IHI, undated; Foster Wheeler, undated, 2011; Babcock and Wilcox, 2011.

APPENDIX B

Supplemental Tables on the Coal Industrial Base Workforce

This appendix provides supplemental tables on the workforce discussed in this monograph.

Table B.1
Workforce Trends in Relevant Occupations for Construction from 2003 to 2009

Occupation	2003	2004	2005	2006	2007	2008	2009
Boilermaker							
National total	20,270	18,520	17,760	17,240	18,650	20,400	22,400
Other heavy and civil engineering construction	290		970	740		190	340
Utility-system construction	560	770		680	1,260		2,600
Carpenter							
National total	852,080	882,490	935,920	985,990	969,670	899,920	743,760
Other heavy and civil engineering construction	4,110	4,880	5,020	6,440	5,500	6,510	5,900
Utility-system construction	6,250	6,460	6,640	6,580	6,850	6,440	6,660
Cement mason or concrete finisher							
National total	180,540	191,690	204,720	218,170	213,850	201,730	165,700
Other heavy and civil engineering construction	1,280	1,470	1,990	2,000	1,800	1,580	1,470
Utility-system construction	2,750	2,940	3,580	4,170	4,330	3,860	3,300
Operating engineer or other construction-equipment operator							
National total	343,640	357,080	378,720	393,090	403,620	398,910	368,200

Table B.1—Continued

Occupation	2003	2004	2005	2006	2007	2008	2009
Other heavy and civil engineering construction	14,120	13,480	14,850	15,590	16,730	16,990	15,450
Utility-system construction	38,140	39,060	41,950	46,110	49,000	49,580	45,010
Electrician							
National total	584,010	582,920	606,500	617,370	624,560	633,010	579,150
Other heavy and civil engineering construction	1,310	1,520	1,230	1,080	1,130	1,150	1,100
Utility-system construction	4,790	4,900	6,030	3,900	4,500	5,160	5,340
Insulation worker, floor, ceiling, and wall							
National total		37,000	34,250	31,450	29,660	28,390	26,500
Other heavy and civil engineering construction		1,030		80			50
Utility-system construction							
Insulation worker, mechanical							
National total		17,110	22,100	27,900	29,110	30,150	29,620
Other heavy and civil engineering construction		540	1,020				
Utility-system construction		260			790	860	1,130

Table B.1—Continued

Occupation	2003	2004	2005	2006	2007	2008	2009
Plumber, pipefitter, or steamfitter							
National total	433,600	424,360	420,770	435,960	435,010	437,540	400,970
Other heavy and civil engineering construction	2,950	3,340	3,200	2,950	1,770	1,250	1,170
Utility-system construction	12,010	11,290	11,950	13,600	15,220	16,100	16,670
Reinforcing iron and rebar worker							
National total	30,250	32,660	30,270	30,180	28,270	28,620	24,200
Other heavy and civil engineering construction		340			170	180	100
Utility-system construction	930	1,000	1,300	630	410	390	360
Millwright							
National total	64,910	57,050	53,080	53,320	49,360	46,250	41,640
Other heavy and civil engineering construction		430	600	480	530	560	420
Utility-system construction	1,160	760	580		1,250	1,380	1,340
Laborer or freight, stock, and material mover, hand							
National total	2,255,780	2,390,910	2,363,960	2,372,130	2,363,440	2,335,510	2,135,790

Table B.1—Continued

Occupation	2003	2004	2005	2006	2007	2008	2009
Other heavy and civil engineering construction	1,710	1,720	1,890	1,890	1,610	1,560	1,550
Utility-system construction	560	770		680	1,260		2,600
Construction manager							
National total	196,110	185,580	192,610	207,630	216,120	220,550	204,760
Other heavy and civil engineering construction	2,850	1,980	2,130	2,190	2,610	2,750	2,960
Utility-system construction	8,910	7,960	8,410	8,670	8,960	9,110	8,760
Engineering manager							
National total	194,940	186,380	187,410	183,960	184,410	182,300	178,110
Other heavy and civil engineering construction	240	200	170	200	320	340	360
Utility-system construction	420	290	470	490	470	400	360
Civil engineer							
National total	206,350	218,220	229,700	236,690	247,370	261,360	259,320
Other heavy and civil engineering construction	1,340	880	820	1,250	2,430		
Utility-system construction	2,330	2,450	1,760	1,610	1,320	1,290	1,470

Table B.1—Continued

Occupation	2003	2004	2005	2006	2007	2008	2009
Civil engineering technician							
National total	90,060	90,000	90,390	86,730	88,030	88,140	82,690
Other heavy and civil engineering construction	190	190	140	110	200	300	330
Utility-system construction	140	240	280	300	160	190	200
Construction laborer							
National total	837,650	854,840	934,000	1,016,530	1,053,060	1,020,290	856,440
Other heavy and civil engineering construction	21,590	18,870	20,670	21,640	23,160	22,410	21,190
Utility-system construction	79,910	75,090	74,760	84,070	87,000	88,890	79,390
Structural iron or steel worker							
National total	70,420	70,240	68,900	67,560	65,100	68,670	65,130
Other heavy and civil engineering construction	690	720	1,060	1,140	1,280	1,190	1,640
Utility-system construction	1,930	2,110	2,580	2,780	2,670	2,250	2,350
Construction or related worker, all other							
National total		81,260	63,340	56,130	58,040	55,820	47,630

Table B.1—Continued

Occupation	2003	2004	2005	2006	2007	2008	2009
Other heavy and civil engineering construction		760	550	540	230		
Utility-system construction		1,120	750	780	950	890	1,220

SOURCE: BLS, 2011.

NOTE: All industry categories are subsets of national data and are mutually exclusive from one another. Blank years represent missing data, not necessarily zero employees.

Table B.2
Workforce Trends in Relevant Occupations for Equipment Manufacturing from 2003 to 2009

Occupation	2003	2004	2005	2006	2007	2008	2009
Boilermaker							
National total	20,270	18,520	17,760	17,240	18,650	20,400	22,400
Boiler, tank, or shipping-container manufacturing	950	880	750	1,150			2,280
Engine, turbine, or power-transmission equipment manufacturing							
Electrician							
National total	584,010	582,920	606,500	617,370	624,560	633,010	579,150
Boiler, tank, or shipping-container manufacturing	520	660	780	870	900	910	800
Engine, turbine, or power-transmission equipment manufacturing	510	1,070	1,240	1,100	860	730	710
Plumber, pipefitter, or steamfitter							
National total	433,600	424,360	420,770	435,960	435,010	437,540	400,970
Boiler, tank, or shipping-container manufacturing	270	80	40	40	210	220	250

Table B.2—Continued

Occupation	2003	2004	2005	2006	2007	2008	2009
Engine, turbine, or power-transmission equipment manufacturing	140	120				70	
Laborer or freight, stock, or material mover, hand							
National total	2,255,780	2,390,910	2,363,960	2,372,130	2,363,440	2,335,510	2,135,790
Boiler, tank, or shipping-container manufacturing	1,520	1,660	1,690	1,650	1,950	1,750	1,600
Engine, turbine, or power-transmission equipment manufacturing	1,050	1,200	1,510	1,410	1,330	940	1,190
Engineering manager							
National total	194,940	186,380	187,410	183,960	184,410	182,300	178,110
Boiler, tank, or shipping-container manufacturing	430	370	370	390	390	410	370
Engine, turbine, or power-transmission equipment manufacturing	1,350	730	720	720	840	980	1,260
General or operations manager							
National total	1,892,060	1,752,910	1,663,810	1,663,280	1,655,410	1,697,690	1,689,680

Table B.2—Continued

Occupation	2003	2004	2005	2006	2007	2008	2009
Boiler, tank, or shipping-container manufacturing	1,510	1,650	1,640	1,800	1,370	1,450	1,390
Engine, turbine, or power-transmission equipment manufacturing	1,110	1,160	1,090	1,270	1,270	1,350	1,190
Industrial production manager							
National total	166,350	155,980	153,950	153,410	152,870	154,030	147,250
Boiler, tank, or shipping-container manufacturing	1,200	1,170	1,030	990	1,060	1,000	970
Engine, turbine, or power-transmission equipment manufacturing	1,140	920	750	850	1,010	1,080	1,010
Industrial engineer							
National total	156,780	174,960	191,640	198,340	204,210	214,580	209,300
Boiler, tank, or shipping-container manufacturing	680	890	1,100	1,270	630	600	550
Engine, turbine, or power-transmission equipment manufacturing	1,370	1,290	1,230		2,030	2,330	1,930

Table B.2—Continued

Occupation	2003	2004	2005	2006	2007	2008	2009
Mechanical drafter							
National total	74,010	76,610	74,650	72,950	74,260	77,070	71,890
Boiler, tank, or shipping-container manufacturing	900	800	910	860	990	970	930
Engine, turbine, or power-transmission equipment manufacturing	810	840	710	660	680	790	900
Industrial engineering technician							
National total	64,260	68,210	73,310	73,640	74,930	72,820	65,460
Boiler, tank, or shipping-container manufacturing	150	100	120	150	100	80	80
Engine, turbine, or power-transmission equipment manufacturing	380	380	420	560	460	640	650
Sheet-metal worker							
National total	189,590	184,740	174,550	177,540	167,730	163,480	146,690
Boiler, tank, or shipping-container manufacturing	1,380	1,580	1,590	1,710	1,740	1,850	1,330

Table B.2—Continued

Occupation	2003	2004	2005	2006	2007	2008	2009
Engine, turbine, or power-transmission equipment manufacturing	100	140	160	90	70	60	90
Industrial machinery mechanic							
National total	192,300	212,770	234,650	250,810	266,550	280,620	276,230
Boiler, tank, or shipping-container manufacturing	620	1,500	1,840	1,970	1,950		2,490
Engine, turbine, or power-transmission equipment manufacturing	1,260	900	1,170	1,460	1,780	1,470	1,190
Assembler or fabricator, all other							
National total		259,830	258,240	288,370	330,940	318,060	267,780
Boiler, tank, or shipping-container manufacturing		480	300	590	880	950	740
Engine, turbine, or power-transmission equipment manufacturing		1,610		3,160	2,320	1,510	1,500
Tool or die maker							
National total	104,210	99,390	99,680	96,960	92,560	85,610	73,640

Table B.2—Continued

Occupation	2003	2004	2005	2006	2007	2008	2009
Boiler, tank, or shipping-container manufacturing	420	470	470	490	490	410	370
Engine, turbine, or power-transmission equipment manufacturing	1,300	1,120	1,060	860	990	850	880
Welder, cutter, solderer, or brazer							
National total	354,300	344,970	358,050	376,630	385,740	392,520	357,740
Boiler, tank, or shipping-container manufacturing	10,630	11,410	12,830	12,790	13,640	14,150	12,860
Engine, turbine, or power-transmission equipment manufacturing	1,730	1,320	1,470	1,480	2,010	1,970	1,730
Welding, soldering, or brazing machine setter, operator, or tender							
National total	53,750	47,210	45,220	48,770	50,820	51,840	41,580
Boiler, tank, or shipping-container manufacturing	2,190	1,930	1,320	1,310	1,300	1,520	1,340
Engine, turbine, or power-transmission equipment manufacturing	330	290	340		490	530	430

Table B.2—Continued

Occupation	2003	2004	2005	2006	2007	2008	2009
Tool grinder, filer, or sharpener							
National total	22,320	19,750	18,180	17,620	17,240	16,410	13,740
Boiler, tank, or shipping-container manufacturing	100	120	140	120	140	140	120
Engine, turbine, or power-transmission equipment manufacturing	590	720	740	460	300	300	290
Coating, painting, or spraying machine setter, operator, or tender							
National total	93,110	96,510	100,830	102,210	102,600	103,310	89,430
Boiler, tank, or shipping-container manufacturing	1,910	2,460	2,620	2,780	2,570	2,810	2,510
Engine, turbine, or power-transmission equipment manufacturing	460	510	390	390	520	520	580
Commercial or industrial designer							
National total	33,390	33,050	31,650	33,540	34,800	32,940	29,170
Boiler, tank, or shipping-container manufacturing	30	270	330	410	90		

Table B.2—Continued

Occupation	2003	2004	2005	2006	2007	2008	2009
Engine, turbine, or power-transmission equipment manufacturing	120		100	160	200	240	170

SOURCE: BLS, 2011.

NOTE: All industry categories are subsets of national data and mutually exclusive from one another. Employees are all part-time and full-time workers who are paid a wage or salary. Blank years represent missing data, not necessarily zero employees. The survey does not cover the self-employed, owners and partners in unincorporated firms, household workers, or unpaid family workers.

Table B.3
Workforce Trends in Relevant Occupations for Operation and Maintenance from 2003 to 2009

Occupation	2003	2004	2005	2006	2007	2008	2009
Boilermaker							
National total	20,270	18,520	17,760	17,240	18,650	20,400	22,400
Utilities					120		190
Carpenter							
National total	852,080	882,490	935,920	985,990	969,670	899,920	743,760
Utilities	260	260	230	200	260	330	
Cement mason or concrete finisher							
National total	180,540	191,690	204,720	218,170	213,850	201,730	165,700
Utilities	30				40	70	
Operating engineer or other construction-equipment operator							
National total	343,640	357,080	378,720	393,090	403,620	398,910	368,200
Utilities	4,000	4,570	4,640	4,800	3,480	3,860	
Electrician							
National total	584,010	582,920	606,500	617,370	624,560	633,010	579,150
Utilities	9,390	9,040	8,130	8,150	8,360	8,510	

Table B.3—Continued

Occupation	2003	2004	2005	2006	2007	2008	2009
Insulation worker, floor, ceiling, or wall							
National total		37,000	34,250	31,450	29,660	28,390	26,500
Utilities		70	90	60			
Insulation worker, mechanical							
National total		17,110	22,100	27,900	29,110	30,150	29,620
Utilities				80	120	120	
Plumber, pipefitter, or steamfitter							
National total	433,600	424,360	420,770	435,960	435,010	437,540	400,970
Utilities	10,010	10,210	9,060	8,460	8,630	8,340	
Reinforcing iron or rebar worker							
National total	30,250	32,660	30,270	30,180	28,270	28,620	24,200
Utilities							
Millwright							
National total	64,910	57,050	53,080	53,320	49,360	46,250	41,640
Utilities	220					90	

Table B.3—Continued

Occupation	2003	2004	2005	2006	2007	2008	2009
Laborer or freight, stock, or material mover, hand							
National total	2,255,780	2,390,910	2,363,960	2,372,130	2,363,440	2,335,510	2,135,790
Utilities	3,330	4,060	3,980	4,340	4,070	3,820	
Construction manager							
National total	196,110	185,580	192,610	207,630	216,120	220,550	204,760
Utilities	560	620	680	580	550	640	770
Engineering manager							
National total	194,940	186,380	187,410	183,960	184,410	182,300	178,110
Utilities	3,930	3,930	3,700	4,220	4,790	4,880	4,630
Civil engineer							
National total	206,350	218,220	229,700	236,690	247,370	261,360	259,320
Utilities	1,810	2,270	2,350	2,300	2,230	2,210	1,890
Civil engineering technician							
National total	90,060	90,000	90,390	86,730	88,030	88,140	82,690
Utilities	1,230	700	780	870	950	900	700

Table B.3—Continued

Occupation	2003	2004	2005	2006	2007	2008	2009
Construction laborer							
National total	837,650	854,840	934,000	1,016,530	1,053,060	1,020,290	856,440
Utilities	1,600	2,420	3,310	3,450	2,960	2,900	
Structural iron or steel worker							
National total	70,420	70,240	68,900	67,560	65,100	68,670	65,130
Utilities		40	60	50			
Construction or related worker, all other							
National total		81,260	63,340	56,130	58,040	55,820	47,630
Utilities		280	270	300	260	320	

SOURCE: BLS, 2011.

NOTE: All industry categories are subsets of national data and mutually exclusive from one another. Employees are all part-time and full-time workers who are paid a wage or salary. The survey does not cover the self-employed, owners and partners in unincorporated firms, household workers, or unpaid family workers. Blank years represent missing data, not necessarily zero employees.

References

ABB, "Alstom Acquires ABB's Share in ABB Alstom Power," press release, March 31, 2000. As of September 21, 2011:
http://www.abb.com/cawp/seitp202/c1256c290031524bc12568b300219d70.aspx

———, "Facts and Figures," edited May 17, 2011. As of October 3, 2011:
http://www.abb.com/cawp/abbzh252/b434095700ab7545c1256ae700494de1.aspx

ADA Environmental Solutions, *Creating a Future with Cleaner Coal: Investor Presentation*, September 2011. As of October 3, 2011:
http://www.adaes.com/PDFs/presentations/Sept%2011%20Investor%20Presentation%20Final.pdf

Alstom, "About Alstom," undated. As of October 3, 2011:
http://www.alstom.com/china/about

American Electric Power, "Ultra-Supercritical Generation," October 2009. As of June 20, 2011:
http://www.aep.com/environmental/climatechange/advancedtechnologies/docs/SuperCriticalFactsheet.pdf

Anderson, David K., and Lewis A. Maroti, "Designing Wet Duct/Stack Systems for Coal-Fired Plants," *Power*, March 15, 2006. As of August 29, 2011:
http://www.powermag.com/environmental/Designing-wet-ductstack-systems-for-coal-fired-plants_536.html

Arena, Mark V., Hans Pung, Cynthia R. Cook, Jefferson P. Marquis, Jessie Riposo, and Gordon T. Lee, *The United Kingdom's Naval Shipbuilding Industrial Base: The Next Fifteen Years*, Santa Monica, Calif.: RAND Corporation, MG-294-MOD, 2005. As of August 29, 2011:
http://www.rand.org/pubs/monographs/MG294.html

Babcock and Wilcox, "Babcock and Wilcox Power Generation Group to Supply Boiler and Environmental Equipment for Wyoming Power Plant," press release, March 27, 2008. As of September 20, 2011:
http://www.babcock.com/news_and_events/2008/20080327a.html

———, *2010 Annual Report*, c. 2011.

BASF, "Key Financial Data," updated July 28, 2011. As of October 3, 2011: http://www.basf.com/group/corporate/en/investor-relations/key-financial-data/index

Basin Electric Power Cooperative, "Dry Fork Station," undated. As of September 20, 2011:
http://www.basinelectric.com/Projects/Dry_Fork_Station/index.html

Bassford, Matt, Hans Pung, Nigel Edgington, Tony Starkey, Kristin Weed, Mark V. Arena, James G. Kallimani, Gordon T. Lee, and Obaid Younossi, *Sustaining Key Skills in the UK Military Aircraft Industry*, Santa Monica, Calif.: RAND Corporation, MG-1023-MOD, 2010. As of August 29, 2011: http://www.rand.org/pubs/monographs/MG1023.html

Bechtel Corporation, "LG&E Selects Bechtel as EPC Contractor on Trimble County #2 Project," press release, March 13, 2006. As of September 20, 2011: http://www.bechtel.com/2006-03-13.html

Bernstein Research, *U.S. Utilities: Coal-Fired Generation Is Squeezed in the Vice of EPA Regulation—Who Wins and Who Loses?* New York, October 2010.

Bipartisan Policy Center, *National Commission on Energy Policy's Task Force on America's Future Energy Jobs*, Washington, D.C., October 1, 2009. As of August 29, 2011:
http://bipartisanpolicy.org/library/report/task-force-americas-future-energy-jobs

Birkler, John, Anthony G. Bower, Jeffrey A. Drezner, Gordon T. Lee, Mark A. Lorell, and Giles K. Smith, *Competition and Innovation in the U.S. Fixed-Wing Military Aircraft Industry*, Santa Monica, Calif.: RAND Corporation, MR-1656-OSD, 2003. As of August 29, 2011:
http://www.rand.org/pubs/monograph_reports/MR1656.html

Black and Veatch, "Black and Veatch–Designed Power Plant Brings 'Economic Boom' and Sustainable Benefits to Northeastern Arkansas," press release, December 15, 2010. As of September 20, 2011:
http://www.bv.com/wcm/press_release/12152010_4981.aspx

BLS—*See* Bureau of Labor Statistics.

Bogner, Ben, "Designing New Composite Stack Liners," *Coal Power*, January 30, 2009. As of August 29, 2011:
http://www.coalpowermag.com/plant_design/Designing-New-Composite-Stack-Liners_183.html

Bradley, Michael J., Susan F. Tierney, Christopher E. Van Atten, and Amlan Saha, *Ensuring a Clean, Modern Electric Generating Fleet While Maintaining Electric System Reliability*, Concord, Mass.: M. J. Bradley and Associates, June 2011. As of September 20, 2011:
http://www.mjbradley.com/sites/default/files/MJBA%20Reliability%20Report%20Update%20June%207%202011.pdf

Bureau of Labor and Statistics, Occupational Employment Statistics program, undated home page. As of March 15, 2011:
http://www.bls.gov/oes/

———, "Download Occupational Employment and Wate Estimates," last modified May 17, 2011; referenced March 15, 2011. As of October 3, 2011:
http://www.bls.gov/oes/oes_dl.htm

Carbon Capture and Sequestration Technologies Program, Massachusetts Institute of Technology, "About the MIT CC&ST Program," undated. As of September 1, 2011:
http://sequestration.mit.edu/index.html

Celebi, Metin, Frank Graves, Gunjan Bathla, and Lucas Bressan, *Potential Coal Plant Retirements Under Emerging Environmental Regulations*, Cambridge, Mass.: Brattle Group, December 8, 2010. As of August 29, 2011:
http://www.brattle.com/_documents/UploadLibrary/Upload898.pdf

Center for Energy Workforce Development, *Gaps in the Energy Workforce Pipeline; A 2007 Workforce Survey Report from the Center for Energy Workforce Development*, c. 2008. As of September 20, 2011:
http://www.cewd.org/surveyreport/cewdreport_oct07.pdf

———, "Gaps in the Energy Workforce Pipeline: 2008 CEWD Survey Results," c. 2009. As of September 20, 2011:
http://www.cewd.org/documents/CEWD_08Results.pdf

———, "Gaps in the Energy Workforce Pipeline: 2009 CEWD Survey Results," c. 2010. As of August 29, 2011:
http://www.cewd.org/mem_resources/
2009%20Survey%20Exec%20Summary.pdf

CEWD—*See* Center for Energy Workforce Development.

Credit Suisse, *Growth from Subtraction*, September 23, 2010. As of September 20, 2011:
http://epw.senate.gov/public/index.cfm?FuseAction=
Files.View&FileStore_id=b42de70d-b814-4410-831d-34b180846a19

D'Olier, Robert, James Bubb, Doug Carroll, Lance Elwell, and Andrew Markel, *DOE NP2010 Nuclear Power Plant Construction Infrastructure Assessment*, Washington, D.C.: U.S. Department of Energy, MPR-2776, revision 0, October 21, 2005. As of August 29, 2011:
http://www.ne.doe.gov/np2010/reports/mpr2776Rev0102105.pdf

Doosan Power Systems, *Building Your Tomorrow Today*, c. 2010. As of October 3, 2011:
http://www.doosan.com/en/documents/annual_brochure/
AnnualReport_english.pdf

———, *All About Power: Annual Report and Accounts 2010*, c. 2011. As of September 20, 2011:
http://www.doosanbabcock.com/live/documents/Doosan_Annual_report_2010.pdf

Drezner, Jeffrey A., Giles K. Smith, Lucille E. Horgan, J. Curt Rogers, and Rachel Schmidt, *Maintaining Future Military Aircraft Design Capability*, Santa Monica, Calif.: RAND Corporation, R-4199-AF, 1992. As of August 29, 2011:
http://www.rand.org/pubs/reports/R4199.html

Duke Energy, "Project Overview," undated. As of March 30, 2011:
http://www.duke-energy.com/about-us/edwardsport-overview.asp

DuPont, *2010 DuPont Data Book*, March 2011.

Electric Power Research Institute, *Program on Technology Innovation: Executive Workshop on the Aging Workforce in the Utility Industry—April 2006, Carnegie Mellon University, Pittsburgh, Pennsylvania: Final Report*, Palo Alto, Calif., 1014815, 2007. As of August 29, 2011:
http://my.epri.com/portal/server.pt?Abstract_id=000000000001014815

Energy Information Administration, "International Energy Statistics," undated. As of August 29, 2011:
http://www.eia.gov/countries/data.cfm

———, *International Energy Outlook 2010*, Washington, D.C., DOE/EIA-0484(2010), July 27, 2010a. As of October 3, 2011:
http://www.eia.gov/forecasts/archive/ieo10/index.html

———, *Annual Energy Review 2009*, Washington, D.C., DOE/EIA-0384(2009), August 19, 2010b. As of August 29, 2011:
http://www.eia.gov/emeu/aer/contents.html

———, "Updated Capital Cost Estimates for Electricity Generation Plants," Washington, D.C., November 2010c. As of August 29, 2011:
http://www.eia.doe.gov/oiaf/beck_plantcosts/index.html

———, *Annual Energy Outlook 2011*, Washington, D.C., DOE/EIA-0383(2011), April 26, 2011a. As of August 29, 2011:
http://www.eia.gov/forecasts/aeo/

———, "Background," last updated May 2011b. As of August 31, 2011:
http://www.eia.gov/countries/cab.cfm?fips=CH

———, *Monthly Energy Review*, Washington, D.C., DOE/EIA-0035(2011/05), May 2011c. As of October 3, 2011:
ftp://ftp.eia.doe.gov/multifuel/mer/00351105.pdf

Foster Wheeler, "Fundamentals: Snapshot," undated.

———, "Foster Wheeler AG Fact Sheet," 2011. As of October 3, 2011:
http://www.fwc.com/corpgov/pdf/factsheet2Q2011729.pdf

Freese, Barbara, Steve Clemmer, Claudio Martinez, and Alan Nogee, *A Risky Proposition: The Financial Hazards of New Investments in Coal Plants*, Cambridge, Mass.: Union of Concerned Scientists, March 2011. As of August 29, 2011: http://www.ucsusa.org/clean_energy/technology_and_impacts/impacts/financial-hazards-of-coal-plant-investments.html

GE, *Growth Starts Here: GE 2010 Annual Report*, c. 2010. As of October 3, 2011: http://www.ge.com/ar2010/pdf/GE_AR10.pdf

Gerdes, Kristin, John Haslbeck, Norma Kuehn, Eric Lewis, Lora L. Pinkerton, Mark Woods, James Simpson, Marc J. Turner, and Elsy Verghese, *Cost and Performance Baseline for Fossil Energy Plants*, Vol. 1: *Bituminous Coal and Natural Gas to Electricity*, Washington, D.C.: National Energy Technology Laboratory, U.S. Department of Energy, DOE/NETL-2010/1397, November 2010. As of August 29, 2011:
http://www.netl.doe.gov/energy-analyses/refshelf/PubDetails.aspx?Action=View&PubId=348

Haldor Topsoe, "Financial Data," undated. As of October 3, 2011: http://www.topsoe.com/about_us/Financial%20data.aspx

Haviland, Richard, senior vice president, construction and major projects, Duke Energy Business Services, on behalf of Duke Energy Indiana, supplemental testimony before the Indiana Utility Regulatory Commission, cause 43114 IGCC 4S1, March 10, 2011.

Hitachi, "Corporate Profile," undated. As of October 3, 2011: http://www.hitachi.com/about/corporate/index.html

IHI, "About IHI," undated. As of October 3, 2011: http://www.ihi.co.jp/en/company/outline/index.html

IHS, *IHS CERA Power Capital Costs Index*, undated; referenced September 6, 2011.

———, "Power Plant Construction Costs: Recovery Paused as Costs Go Flat Once More," press release, Cambridge, Mass., December 21, 2010. As of August 29, 2011:
http://press.ihs.com/press-release/energy-power/power-plant-construction-costs-recovery-paused-costs-go-flat-once-more

———, "Power Plant Construction Costs: Cost Pressures Returning," press release, July 6, 2011. As of September 20, 2011:
http://press.ihs.com/press-release/energy-power/power-plant-construction-costs-cost-pressures-returning

Industcards, "Power Plants Around the World," undated home page. As of March 11, 2011:
http://www.industcards.com/

Interagency Task Force on Carbon Capture and Storage, *Report of the Interagency Task Force on Carbon Capture and Storage*, August 2010. As of September 20, 2011: http://fossil.energy.gov/programs/sequestration/ccstf/CCSTaskForceReport2010.pdf

Kitto, John B., and S. C. Stultz, eds., *Steam, Its Generation and Use*, 41st ed., Barberton, Ohio: Babcock and Wilcox, 2005.

Macedonia, Jennifer, Joe Kruger, Lourdes Long, and Meghan McGuinness, *Environmental Regulation and Electric System Reliability*, Washington, D.C.: Bipartisan Policy Center, June 13, 2011. As of September 20, 2011: http://www.bipartisanpolicy.org/library/report/environmental-regulation-and-electric-system-reliability

McDermott International, *Annual Report*, 2010.

Merrow, Edward W., *Cost Growth in New Process Facilities*, Santa Monica, Calif.: RAND Corporation, P-6869, 1983. As of August 29, 2011: http://www.rand.org/pubs/papers/P6869.html

———, *An Analysis of Cost Improvement in Chemical Process Technologies*, Santa Monica, Calif.: RAND Corporation, R-3357-DOE, 1989. As of August 29, 2011: http://www.rand.org/pubs/reports/R3357.html

Merrow, Edward W., Stephen W. Chapel, and Christopher Worthing, *A Review of Cost Estimation in New Technologies: Implications for Energy Process Plants*, Santa Monica, Calif.: RAND Corporation, R-2481-DOE, 1979. As of August 29, 2011: http://www.rand.org/pubs/reports/R2481.html

Merrow, Edward W., Lorraine M. McDonnell, and R. Ylmaz Arguden, *Understanding the Outcomes of Mega-Projects: A Quantitative Analysis of Very Large Civilian Projects*, Santa Monica, Calif.: RAND Corporation, R-3560-PSSP, 1988. As of August 29, 2011: http://www.rand.org/pubs/reports/R3560.html

Merrow, Edward W., Kenneth Phillips, and Christopher W. Myers, *Understanding Cost Growth and Performance Shortfalls in Pioneer Process Plants*, Santa Monica, Calif.: RAND Corporation, R-2569-DOE, 1981. As of August 29, 2011: http://www.rand.org/pubs/reports/R2569.html

Metso, "Power," undated. As of September 21, 2011: http://www.metso.com/corporation/about_eng.nsf/WebWID/WTB-090520-2256F-1F7EB?OpenDocument

———, "Metso in Brief," updated September 7, 2011. As of October 3, 2011: http://www.metso.com/corporation/about_eng.nsf/WebWID/WTB-041026-2256F-55957?OpenDocument

NAES Corporation, "NAES Corporation Operating 665 MW Coal Plant," press release, January 6, 2011. As of September 20, 2011: http://www.naes.com/news/naes-corporation-operating-665-mw-coal-plant

National Energy Technology Laboratory, "Carbon Sequestration: Program Overview," undated. As of September 20, 2011:
http://www.netl.doe.gov/technologies/carbon_seq/overview.html

———, *2007 Coal Power Plant DataBase*, 2007. As of September 21, 2011:
http://www.netl.doe.gov/energy-analyses/hold/technology.html

Ochs Center for Metropolitan Studies, *A Fraction of the Jobs: A Case Study of the Job Creation Impact of Completed Coal-Fired Power Plants Between 2005 and 2009*, April 29, 2011. As of September 20, 2011:
http://www.afractionofthejobs.com/pdf/
the-Ochs-Center-A-Fraction-of-the-Jobs.pdf

Ortiz, David S., Aimee E. Curtright, Constantine Samaras, Aviva Litovitz, and Nicholas Burger, *Near-Term Opportunities for Integrating Biomass into the U.S. Electricity Supply: Technical Considerations*, Santa Monica, Calif.: RAND Corporation, TR-984-NETL, 2011. As of September 20, 2011:
http://www.rand.org/pubs/technical_reports/TR984.html

Peltier, Robert, "J. K. Spruce Power Plant, Unit 1, San Antonio, Texas," *Power*, October 15, 2008. As of September 20, 2011:
http://www.powermag.com/coal/1432.html

———, "Plant of the Year: KCP&L's Iatan 2 Earns Power's Highest Honor," *Power*, August 1, 2011. As of September 20, 2011:
http://www.powermag.com/issues/cover_stories/
Plant-of-the-Year-KCP-and-Ls-Iatan-2-Earns-POWERs-Highest-Honor_3882.html

Prairie State Energy Campus, "Prairie State Energy Campus: December 2010," December 2010. As of August 29, 2011:
http://www.prairiestateenergycampus.com/imagesuploaded//
2010-12_PSEC_Presentation.pdf

———, "March 2011," c. 2011. As of September 1, 2011:
http://www.prairiestateenergycampus.com/
pages.asp?pagemainlevel=6&pageid=142

Prevost, Lisa, "Steps-from-Work Housing," *New York Times*, August 19, 2011. As of September 21, 2011:
http://www.nytimes.com/2011/08/21/realestate/
steps-from-work-housing-in-the-regionconnecticut.html

Price, Jason, Nadav Tanners, Jim Neumann, and Roy Oommen, *Employment Impacts Associated with the Manufacture, Installation, and Operation of Scrubbers*, Industrial Economics, March 31, 2011. As of September 20, 2011:
http://www.epa.gov/ttn/atw/utility/scrubber_jobs_memo_3-31-11.pdf

Public Law 88-206, Clean Air Act, 1963.

Public Law 101-549, Clean Air Act Amendments, November 15, 1990.

Pung, Hans, Laurence Smallman, Mark V. Arena, James G. Kallimani, Gordon T. Lee, Samir Puri, and John F. Schank, *Sustaining Key Skills in the UK Naval Industry*, Santa Monica, Calif.: RAND Corporation, MG-725-MOD, 2008. As of August 30, 2011:
http://www.rand.org/pubs/monographs/MG725.html

Schank, John F., Mark V. Arena, Paul DeLuca, Jessie Riposo, Kimberly Curry Hall, Todd Weeks, and James Chiesa, *Sustaining U.S. Nuclear Submarine Design Capabilities*, Santa Monica, Calif.: RAND Corporation, MG-608-NAVY, 2007. As of August 30, 2011:
http://www.rand.org/pubs/monographs/MG608.html

Schank, John F., Jessie Riposo, John Birkler, and James Chiesa, *The United Kingdom's Nuclear Submarine Industrial Base*, Vol. 1: *Sustaining Design and Production Resources*, Santa Monica, Calif.: RAND Corporation, MG-326/1-MOD, 2005. As of August 30, 2011:
http://www.rand.org/pubs/monographs/MG326z1.html

Schlissel, David, Allison Smith, and Rachel Wilson, *Coal-Fired Power Plant Construction Costs*, Cambridge, Mass.: Synapse Energy Economics, July 2008. As of August 30, 2011:
http://www.synapse-energy.com/Downloads/
SynapsePaper.2008-07.0.Coal-Plant-Construction-Costs.A0021.pdf

Seong, Somi, Obaid Younossi, Benjamin W. Goldsmith, Thomas Lang, and Michael J. Neumann, *Titanium: Industrial Base, Price Trends, and Technology Initiatives*, Santa Monica, Calif.: RAND Corporation, MG-789-AF, 2009. As of September 20, 2011:
http://www.rand.org/pubs/monographs/MG789.html

Shuster, Erik, Office of Strategic Energy Analysis and Planning, National Energy Technology Laboratory, "Tracking New Coal-Fired Power Plants," briefing, January 14, 2011. As of August 29, 2011:
http://www.netl.doe.gov/coal/refshelf/ncp.pdf

Siemens, "Siemens at a Glance," undated. As of October 3, 2011:
http://www.siemens.com/investor/en/company_overview.htm

"Southwest Power Station Unit 2 for City Utilities of Springfield, Missouri, USA," *Power-Technology.com*, undated. As of September 20, 2011:
http://www.power-technology.com/projects/springfield/

Spring, Nancy, "Supercritical Plants to Come Online in 2009," *Power Engineering*, July 2009, pp. 64–71. As of September 20, 2011:
http://online.qmags.com/PE0709/
Default.aspx?sessionID=8DC502B9BECC1C8A0C476B00B&cid=969220&
eid=13991

Sun, Guodong, *Coal in China: Resources, Uses, and Advanced Coal Technologies*, Arlington, Va.: Pew Center on Global Climate Change, March 2010. As of August 30, 2011:
http://www.pewclimate.org/white-papers/coal-initiative/coal-china-resources-uses-technologies

Tampa Electric Company, Polk Power Station, *Tampa Electric Polk Power Station Integrated Gasification Combined Cycle Project: Final Technical Report*, Morgantown, W.Va.: U.S. Department of Energy, Office of Fossil Energy, National Energy Technology Laboratory, August 2002. As of August 29, 2011:
http://www.tampaelectric.com/data/files/PolkDOEFinalTechnicalReport.pdf

"The Top 200 Environmental Firms," *Engineering News-Record*, 2011. As of October 3, 2011:
http://enr.construction.com/toplists/EnvironmentalFirms/001-100.asp

"The Top 200 International Design Firms," *Engineering News-Record*, 2011. As of October 3, 2011:
http://enr.construction.com/toplists/InternationalDesignFirms/001-100.asp

"The Top 225 International Contractors," *Engineering News-Record*, 2011. As of October 3, 2011:
http://enr.construction.com/toplists/InternationalContractors/001-100.asp

"The Top 400 Contractors," *Engineering News-Record*, 2011. As of September 20, 2011:
http://enr.construction.com/toplists/contractors/001-100.asp

"The Top 500 Design Firms," *Engineering News-Record*, 2011. As of September 20, 2011:
http://enr.construction.com/toplists/designfirms/001-100.asp

Toshiba, "Corporate Data," undated. As of October 3, 2011:
http://www.toshiba.co.jp/worldwide/about/corp_data.html

———, *Annual Report: Operational Review 2011*, c. 2011. As of October 3, 2011:
http://www.toshiba.co.jp/about/ir/en/library/ar/ar2011/tar2011e.pdf

U.S. Department of Agriculture, *Draft Environmental Impact Statement for the Basin Electric Power Cooperative: Environmental Impact Statement on Dry Fork Station and Hughes Transmission Line*, Washington, D.C., August 2007. As of August 30, 2011:
http://www.rurdev.usda.gov/UWP-Dry-Fork-Station.html

U.S. Department of Energy, *Workforce Trends in the Electric Utility Industry: A Report to the United States Congress Pursuant to Section 1101 of the Energy Policy Act of 2005*, August 2006. As of September 20, 2011:
http://energy.gov/sites/prod/files/oeprod/DocumentsandMedia/Workforce_Trends_Report_090706_FINAL.pdf

U.S. Department of Labor, Employment and Training Administration, *Identifying and Addressing Workforce Challenges in America's Energy Industry*, March 2007. As of September 20, 2011:
http://www.doleta.gov/BRG/pdf/Energy%20Report_final.pdf

U.S. Environmental Protection Agency, "Hazardous and Solid Waste Management System; Identification and Listing of Special Wastes; Disposal of Coal Combustion Residuals From Electric Utilities; Proposed Rule," *Federal Register*, Vol. 75, No. 118, June 21, 2010, pp. 35127–35264. As of September 20, 2011:
http://www.regulations.gov/
#!documentDetail;D=EPA-HQ-RCRA-2009-0640-0352

———, "Clean Air Act," last updated March 1, 2011a. As of June 24, 2011:
http://www.epa.gov/air/caa/

———, *An Assessment of the Feasibility of Retrofits for the Toxics Rule*, March 9, 2011b. As of September 20, 2011:
http://www.epa.gov/ttn/atw/utility/pro/feasibility_retrofit.pdf

———, "National Pollutant Discharge Elimination System: Cooling Water Intake Structures at Existing Facilities and Phase I Facilities," *Federal Register*, Vol. 76, No. 76, April 20, 2011c, pp. 22174–22288. As of September 20, 2011:
http://www.gpo.gov/fdsys/pkg/FR-2011-04-20/pdf/2011-8033.pdf

———, "National Emission Standards for Hazardous Air Pollutants From Coal- and Oil-Fired Electric Utility Steam Generating Units and Standards of Performance for Fossil-Fuel-Fired Electric Utility, Industrial-Commercial-Institutional, and Small Industrial-Commercial-Institutional Steam Generating Units; Proposed Rule," *Federal Register*, Vol. 76, No. 85, May 3, 2011d, pp. 24976–25147. As of September 20, 2011:
http://www.epa.gov/ttn/atw/utility/fr03my11.pdf

———, "Addressing Greenhouse Gas Emissions," last updated June 14, 2011e. As of September 20, 2011:
http://www.epa.gov/airquality/ghgsettlement.html

———, "Federal Implementation Plans: Interstate Transport of Fine Particulate Matter and Ozone and Correction of SIP Approvals," *Federal Register*, Vol. 76, No. 152, August 8, 2011f, pp. 48208–48483. As of September 1, 2011:
http://www.gpo.gov/fdsys/pkg/FR-2011-08-08/pdf/2011-17600.pdf

USDA—*See* U.S. Department of Agriculture.

Ventyx, *Velocity Suite EV Power Database*, 2011; referenced March 3, 2011.

"Wickes Companies, Inc.," *International Directory of Company Histories*, 1992. As of September 21, 2011:
http://www.encyclopedia.com/doc/1G2-2840900088.html

Wood Mackenzie, "Wood Mackenzie: Long-Term Viability of Many US Coal Plants at Risk," press release, Houston, Texas, September 17, 2010. As of August 30, 2011:
http://www.woodmacresearch.com/cgi-bin/corp/portal/corp/corpPressDetail.jsp?oid=2178098